M000268629

CHRISTIAN LIGHT PUBLICATIONS INC.
P.O. BOX 1212
Harrisonburg, Virginia 22803-1212
(540) 434-0768

Exploring Agriscience

3rd Edition

Thomson Delmar Learning
is proud to
support
FFA Activities

Exploring Agriscience

3rd Edition

RAY V. HERREN

THOMSON

DELMAR LEARNING ™

Australia Canada Mexico Singapore Spain United Kingdom United States

THOMSON

DELMAR LEARNING

Exploring Agriscience, 3rd Edition
by Ray V. Herren

Vice President, Career Education Strategic Business Unit:
Dawn Gerrain

Director of Editorial:
Sherry Gomoll

Acquisitions Editor:
David Rosenbaum

Developmental Editor:
Gerald O'Malley

Editorial Assistant:
Christina Gifford

Director of Production:
Wendy A. Troeger

Production Manager:
JP Henkel

Production Editor:
Matthew Williams

Technology Project Manager:
Sandy Charette

Director of Marketing:
Wendy E. Mapstone

Marketing Specialist:
Gerard McAvey

Cover Designer:
Suzanne Nelson

Library of Congress Cataloging-in-Publication Data

Herren, Ray V.
   Exploring agriscience / Ray V. Herren.— 3rd ed.
      p. cm.
   Includes index.
   ISBN 13: 978-1-4018-9644-7
   ISBN 10: 1-4018-9644-8
   1. Agriculture—Juvenile literature. I. Title.
   S495 .H62 2006
   630—dc22   2005007714

NOTICE TO THE READER

Contents

Preface

Agriculture is a rapidly changing science. Change comes about so rapidly that it is difficult to keep up with all the developments. However, basic scientific principles are the foundation of modern agriculture and these principles drive change. *Exploring Agriscience* Third Edition is meant to be the first text that introduces students to the dynamic industry of agriculture. Integrated throughout are principles of science behind the industry. Research has shown that most Americans have a misconception of agriculture. Many think of agriculture as farming. While this is a basic component of agriculture, the scope of agriculture is much broader. This industry employs more people than any other and is responsible for much of the nation's wealth. This text provides students with an overview of the different aspects of that industry. Of course, all aspects of agriculture are not covered. To address all phases of the industry would take many volumes and would be beyond the confines of classroom teaching. Rather, this text introduces the main areas of agriculture that have made the industry great.

The third edition of *Exploring Agriscience* contains four new chapters. Chapter 19 explores the science of food preservation. Chapter 21, Selecting and Using Hand Tools, begins a new focus in Agriscience—that of the science of physics. The next chapter, Small Engine Operation, delves into the study of power mechanics. The study of the mechanical aspects of agriculture is as important as the biological side. Mechanization has allowed the discoveries in biology to be used to their potential. The fourth new chapter deals with high school agriculture programs and is intended to stimulate interest in continuing the study of agriculture.

All aspects of the agricultural industry are based on science. Principles of biology and physics have been researched, manipulated, and used to provide advancements in the production of food and fiber. These advancements have made life better for all Americans. No education would be complete without an understanding of the role of agriculture in our lives.

This book was written with the hope that students will continue the study of the science of agriculture. Within each chapter are highlights of particular FFA activities for high school students. These are by no means the only programs of the organization. They represent areas that coincide with the content area of each chapter. Any student with an interest in biological science will find a program in Agriculture Education and the FFA that will match that interest.

The addition of the chapter on high school agriculture programs will provide students with more insights into the program.

This new and exciting title from Thomson Delmar Learning is written to be used as a core text for a middle school agriscience curriculum.

Features

▼ Full color photos, illustrations, and design stimulate student interest and make difficult concepts easier for student comprehension.

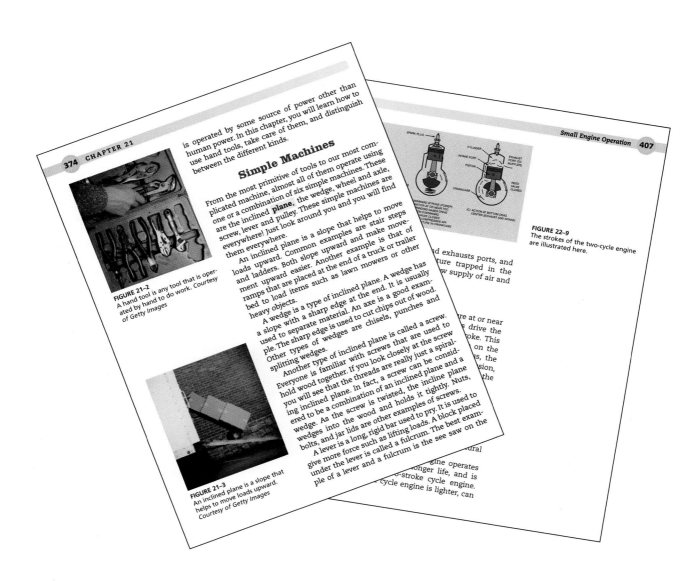

374 CHAPTER 21

is operated by some source of power other than human power. In this chapter, you will learn how to use hand tools, take care of them, and distinguish between the different kinds.

Simple Machines

From the most primitive of tools to our most complicated machine, almost all of them operate using one or a combination of six simple machines. These are the inclined **plane**, the wedge, wheel and axle, screw, lever and pulley. These simple machines are everywhere! Just look around you and you will find them everywhere.

An inclined plane is a slope that helps to move loads upward. Common examples are stair steps and ladders. Both slope upward and make movement upward easier. Another example is that of ramps that are placed at the end of a truck or trailer bed to load items such as lawn mowers or other heavy objects.

A wedge is a type of inclined plane. A wedge has a slope with a sharp edge at the end. It is usually used to separate material. An axe is a good example. The sharp edge is used to cut chips out of wood. Other types of wedges are chisels, punches and splitting wedges.

Another type of inclined plane is called a screw. Everyone is familiar with screws that are used to hold wood together. If you look closely at the screw you will see that the threads are really just a spiraling inclined plane. In fact, a screw can be considered to be a combination of an inclined plane and a wedge. As the screw is twisted, the incline plane wedges into the wood and holds it tightly. Nuts, bolts, and jar lids are other examples of screws.

A lever is a long, rigid bar used to pry. It is used to give more force such as lifting loads. A block placed under the lever is called a fulcrum. The best example of a lever and a fulcrum is the see saw on the

FIGURE 21–2
A hand tool is any tool that is operated by hand to do work. Courtesy of Getty Images

FIGURE 21–3
An inclined plane is a slope that helps to move loads upward. Courtesy of Getty Images

Small Engine Operation 407

FIGURE 22–9
The strokes of the two-cycle engine are illustrated here.

▼ Boxed articles entitled "Career Development Events" feature FFA activities and competitions.

▼ Numerous hands-on activities and discussion questions reinforce learning of material and concepts covered in the text.

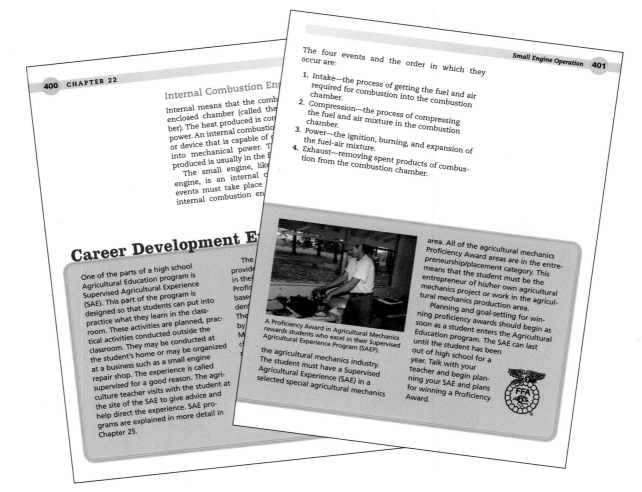

Internal Combustion En[g]

Internal means that the comb[ustion]
enclosed chamber (called the [cham-]
ber). The heat produced is con[verted to]
power. An internal combustio[n engine]
or device that is capable of [converting]
into mechanical power. T[he power]
produced is usually in the f[orm]

The small engine, like [the larger]
engine, is an internal c[ombustion]
events must take place [in an]
internal combustion en[gine.]

The four events and the order in which they occur are:

1. Intake—the process of getting the fuel and air required for combustion into the combustion chamber.
2. Compression—the process of compressing the fuel and air mixture in the combustion chamber.
3. Power—the ignition, burning, and expansion of the fuel-air mixture.
4. Exhaust—removing spent products of combustion from the combustion chamber.

Career Development E[vents]

One of the parts of a high school Agricultural Education program is Supervised Agricultural Experience (SAE). This part of the program is designed so that students can put into practice what they learn in the classroom. These activities are planned, practical activities conducted outside the classroom. They may be conducted at the student's home or may be organized at a business such as a small engine repair shop. The experience is called *supervised* for a good reason. The agriculture teacher visits with the student at the site of the SAE to give advice and help direct the experience. SAE programs are explained in more detail in Chapter 25.

The [program also]
provide[s]
in the [form of]
Profic[iency]
base[d]
den[t]
The [award is sponsored]
by [the]
M[onsanto]

the agricultural mechanics industry. The student must have a Supervised Agricultural Experience (SAE) in a selected special agricultural mechanics

A Proficiency Award in Agricultural Mechanics rewards students who excel in their Supervised Agricultural Experience Program (SAEP).

area. All of the agricultural mechanics Proficiency Award areas are in the entrepreneurship/placement category. This means that the student must be the entrepreneur of his/her own agricultural mechanics project or work in the agricultural mechanics production area.

Planning and goal-setting for winning proficiency awards should begin as soon as a student enters the Agricultural Education program. The SAE can last until the student has been out of high school for a year. Talk with your teacher and begin planning your SAE and plans for winning a Proficiency Award.

Enhanced Content

▼ Provides a hands-on exploratory, science-based approach to agriscience designed to be used as a student's first text in agriculture.

▼ Broad-based coverage of horticulture, animal science, environmental science, and biotechnology is presented in an easy-to-understand format for middle school students.

▼ Four new chapters (Food Preservation, Hand Tools, Small Engines, and High School Agriculture Programs) reflect Thomson Delmar Learning's commitment to bringing the latest trends in agriculture education to the classroom.

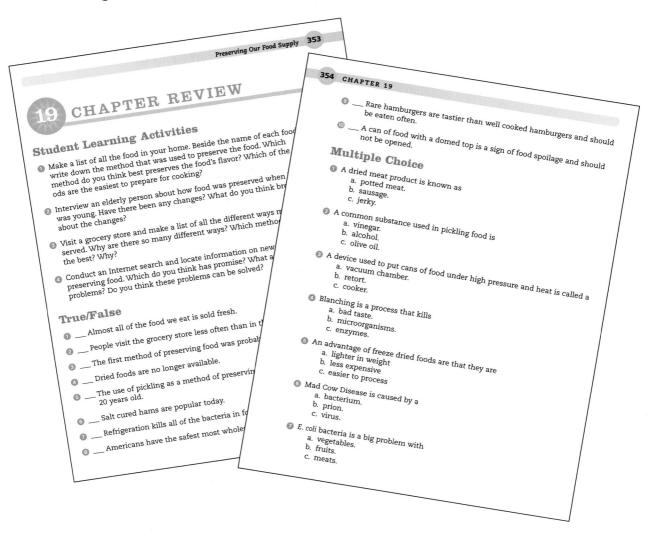

Preserving Our Food Supply   353

19 CHAPTER REVIEW

Student Learning Activities

1. Make a list of all the food in your home. Beside the name of each food write down the method that was used to preserve the food. Which method do you think best preserves the food's flavor? Which of the foods are the easiest to prepare for cooking?

2. Interview an elderly person about how food was preserved when was young. Have there been any changes? What do you think br about the changes?

3. Visit a grocery store and make a list of all the different ways served. Why are there so many different ways? Which metho the best? Why?

4. Conduct an Internet search and locate information on new preserving food. Which do you think has promise? What a problems? Do you think these problems can be solved?

True/False

1. ___ Almost all of the food we eat is sold fresh.

2. ___ People visit the grocery store less often than in t

3. ___ The first method of preserving food was probab

4. ___ Dried foods are no longer available.

5. ___ The use of pickling as a method of preservin 20 years old.

6. ___ Salt cured hams are popular today.

7. ___ Refrigeration kills all of the bacteria in fo

8. ___ Americans have the safest most whole

---

354 CHAPTER 19

9. ___ Rare hamburgers are tastier than well cooked hamburgers and should be eaten often.

10. ___ A can of food with a domed top is a sign of food spoilage and should not be opened.

Multiple Choice

1. A dried meat product is known as
   a. potted meat.
   b. sausage.
   c. jerky.

2. A common substance used in pickling food is
   a. vinegar.
   b. alcohol.
   c. olive oil.

3. A device used to put cans of food under high pressure and heat is called a
   a. vacuum chamber.
   b. retort.
   c. cooker.

4. Blanching is a process that kills
   a. bad taste.
   b. microorganisms.
   c. enzymes.

5. An advantage of freeze dried foods are that they are
   a. lighter in weight
   b. less expensive
   c. easier to process

6. Mad Cow Disease is caused by a
   a. bacterium.
   b. prion.
   c. virus.

7. E. coli bacteria is a big problem with
   a. vegetables.
   b. fruits.
   c. meats.

Extensive Teaching/ Learning Package

The complete supplement package was developed to achieve two goals:

1. To assist students in learning the essential information needed to continue their exploration into the exciting field of agriscience.

2. To assist instructors in planning and implementing their instructional program for the most efficient use of time and other resources.

Instructor's Guide to Text

The Instructor's Guide provides answers to the end-of-chapter questions and additional material to assist the instructor in the preparation of lesson plans.

Lab Manual

Order #1-4018-9647-2

This comprehensive lab manual reinforces the text content. It is recommended that students complete each lab to confirm understanding of essential science content. Great care has been taken to provide instructors with low cost, strong content labs.

Lab Manual CD-ROM

Order #1-4018–9649–9

Our comprehensive student lab manual is now available on a CD-ROM. Instructor's now have the capability of printing off as many copies of the exercises as needed, for as long as they own the book.

Lab Manual Instructor's Guide

The Instructor's Guide provides answers to lab manual exercises and additional guidance for the instructor.

Classmaster CD-ROM

Order #1–4018–9650–2

This supplement provides the instructor with valuable resources to simplify the planning and implementation of the instructional program. It includes answers to questions in the text, lesson plans, a computerized test bank, and PowerPoint Presentation slides to provide the instructor with a cohesive plan for presenting each topic.

Classroom Interactivity CD-ROM

Order #1-4018-9652-9

This new supplement provides instructor's the opportunity to create a dynamic learning environment while engaging their students in active participation. This CD-ROM contains four different "game-show-themed" applications to be run by the instructor. All questions are taken directly from the readings of the text and serve as great tools in reinforcing the main concepts of each unit.

Acknowledgments

Mary Herren for her help in proofing, editing, and assembling materials. Valery Moseley for her help with photos and providing the chapter on landscaping.

The author and Delmar would like to thank the reviewers who shared their content expertise and added valuable suggestions for this second edition:

Maxine Chipman
Wonewoc-Center Schools
Wonewoc, Wisconsin

Odon Russell
Paxton High School
Paxton, Florida

Lynn M. Skutches
Summerville High School
Sonora, California

Perry Richard
Harmony Grove High School
Camden, Arizona

Dedication

This book is dedicated to my Aunt Flossie Herren. It was she who gave me my first book and kept me supplied with material to read. She dedicated her life to teaching children and was a constant source of encouragement to me.

The World of Agriculture

Student Objectives

When you have finished studying this chapter, you should be able to:

- Discuss why agriculture is important in our everyday lives.
- Explain why agriculture is such a broad and diverse industry.
- Explain why agriculture is a science.
- Summarize the reasons why agriculture contributes to the wealth of the nation.
- List and discuss the uses of plants.
- List and discuss the uses of animals.
- Contrast agricultural exports and imports.
- Discuss the processing and distribution of agricultural products.

Key Terms

agriculture	export	growing season
biology	import	
ecology	commodities	

You are about to begin a study of one of the most dynamic industries the world has ever known! This industry—American **agriculture**—is involved with growing plants and animals for use by all the people of this country and much of the world. We live in a country with one of the highest standards of living in the world. Americans are the envy of most of the world! Much of the affluence we enjoy is either a direct or indirect result of the agriculture industry. Compared to people in other countries, we pay a very small portion of our income for food. We spend around 12 to 15 percent of our income on food (Figure 1–1). This compares to over 70 percent in some countries. Our prosperity began because of the fertile soil and ideal climate we enjoy in the United States. Our soil and climate provided the basis for our great agricultural industry. New discoveries and developments as a result of scientific research in agriculture promoted the growth of the agriculture industry. This industry has been one of the major causes of the wealth of the country.

FIGURE 1–1
Americans only spend about 12 to 15 percent of their income for food. *Courtesy of Getty Images.*

Science and Agriculture

Science is the study of or the explanation of natural phenomena. In other words, science deals with the laws of nature and our understanding of them. Agriculture fits this definition very well. No other discipline deals more directly with the laws of nature than does agriculture. Recognized areas of science such as physics, chemistry, geology, meteorology, and **biology** all are a part of agriculture. Biology is the study of living organisms—their life cycles and how they live, grow, and reproduce. All of the knowledge we have of living things (biology) has only three applications: medicine, **ecology**, and agriculture. In fact, medicine and ecology (the study of the relationship of organisms in an environment) are related to agriculture. Medicine is related to agriculture because good nutrition is dependent on humans obtaining plenty of wholesome food. Ecology is related because of the tremendous impact agriculture has on the environment.

Agriculture is the most important of all the sciences because we depend on agriculture for the food we eat. Without a steady intake of food we would soon die. Through the scientific developments that have made agriculture so efficient, fewer than half of the people in this country are involved with getting food to the rest of the population. Our system has become so efficient that one American farmer feeds more than 100 people (Figure 1–2). Of course, a large number of people support the farmer in providing supplies and distributing the produce. This allows all of the people not involved in agriculture to produce manufactured goods and conduct services that would otherwise be impossible if it were not for our agricultural system, people would spend most of their time gathering something to eat.

FIGURE 1–2
One American farmer feeds
more than 100 people.
Courtesy of Getty Images.

Many people think of agriculture in terms of a farmer out plowing a field, and this is a part of agriculture. However, farmers represent a small percentage of the people who work in agriculture. The agriculture industry hires more people each year than any other industry (Figure 1–3). These jobs involve such diverse occupations as marketing specialists, chemical manufacturers, food inspectors, and research scientists. Yet all of these people are a part of agriculture.

The growing of plants and animals involves many different segments. At one time, the growing of living things was only for food and clothing. In modern times in this country, many other uses have been made of the plants and animals we grow. In fact new uses are constantly being found. For example, some scientists think that in the future we will be able to grow plants that will supply the fuel needed to run our automobiles!

FIGURE 1–3
The industry of agriculture hires more people than any other industry. *Courtesy of James Strawser, The University of Georgia.*

Growing Plants

In a very real sense, the foundation of all agriculture is the growing of plants. Without plants, the animals we grow for food would have nothing to

FIGURE 1–4
Agriculture supplies us with nutritious food. *Courtesy of Getty Images.*

eat. Much of the food we eat comes from plants (Figure 1–4). The flour that goes into our breads and the oil that we use for cooking come from plants. The fresh fruits that supply us with vitamins and other nutrients are plant products. Just think of all the vegetables you ate during the past week! All of these were produced by our agriculture industry. The seeds of plants supply us with a rich source of protein. Did you know that peanut butter is made from the roasted and ground-up seeds of a plant?

Plants also supply us with clothing. The fiber from cotton plants gives us comfortable clothing. This crop has been grown in this country since before it was a country. Today, only China grows more cotton than the United States.

Our houses also come from plants. Trees supply the lumber for buildings and furniture. You might not have thought of forestry as being agriculture, but most of the trees that are harvested each year have been planted and cared for by tree producers (Figure 1–5). Trees also supplied the paper to make this book. Just think of all the paper Americans use every day. This industry represents a large portion of the economy of the nation.

FIGURE 1–5
The growing of trees is a part of agriculture. *Courtesy of James Strawser, The University of Georgia.*

Many medicines come from plants. A good example is the drug digitalis that is used to help people with heart problems. This important medicine comes from the foxglove plant, which is grown by agricultural producers (Figure 1–6).

One large and expanding aspect of agriculture is the growing of ornamental plants. Americans spend millions of dollars each year buying plants to beautify their homes. Drive through almost any neighborhood and you will see a broad variety of plants used to enhance the value of homes. Trees, shrubs, grass seed, and turf are all grown on farms and nurseries. Potted plants and flowers are produced in greenhouses for people to buy. Each year, many people make their living raising plants in greenhouses.

Plants are used in ways you probably never even considered. Did you realize that chewing gum comes almost entirely from plants? The gum is made from a substance called chicle that comes from the sap of a tree. The sugar used to sweeten it comes from sugar beets or sugarcane (Figure 1–7). Even flavorings such as mint come from cultivated plants. All these plants are planted, cultivated, and harvested to produce chewing gum. This makes chewing gum an agricultural product!

FIGURE 1–6
Plants like this foxglove plant provide us with medicines. *Courtesy of James Strawser, The University of Georgia.*

FIGURE 1–7
The sugar we use comes from crops such as sugarcane. *Courtesy of John Wasniak, Louisiana Agricultural Experiment Station.*

The growing of the plants is supported by other parts of agriculture. Fertilizers are produced that supply the plants with nutrients. Chemicals are manufactured to produce pesticides that kill weeds and insects that attack the plants. Tractors, cultivators, and harvesting equipment have to be designed and built (Figure 1–8). Fuel and lubricants have to be produced for the tractors and equipment. Large irrigation systems have to be built, installed, and maintained. All of these items have to be marketed and sold to the producer. The growing of plants is big business!

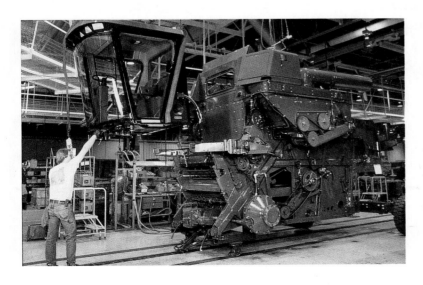

FIGURE 1–8
All of the machinery used in agriculture has to be designed and built. *Courtesy of John Deere Company.*

Career Development Events

When you enroll in high school agriculture education courses, you become a member of the Future Farmers of America (FFA). This organization gives you opportunities you may have never even dreamed possible. The skills and concepts you learn in the classroom and FFA activities are a fun way to learn. You can even travel to other countries to learn about agriculture.

The first FFA International Exchange Program was with England in 1948. Since then the FFA international program has grown along with the realization that a world food supply is an important part of the agriculture in the United States. FFA international activities are designed to link FFA members with other nations to provide experiences that will improve agriculture education around the globe. There are several ways in which an FFA member can become involved in an international exchange program.

In your local FFA school chapter, you can be part of the international program by promoting and conducting FFA development projects. The development projects can involve single FFA members and chapters nationwide. The FFA can help by improving agriculture practices and crops in developing countries, helping world hunger problems, establishing rural agriculture education programs, and developing of new technology in agriculture. Projects for FFA chapters range from raising funds to support projects to purchasing books, tools, or seed for a rural agriculture education facility.

The chapter may even sponsor and house an exchange student from another country. The World Agriscience Studies (WASS) is an international agricultural student exchange program. Students from other nations are placed with American host families to study agriculture in the United States. The local FFA chapter can provide foreign students with experiences in agriculture by involving them in their local

The Animal Industry

Most of the animals produced in this country are raised for food. The cattle industry supplies us with beef. The steaks you eat for dinner and the hamburgers you buy at the fast food restaurant come

supervised agriculture experience projects. FFA members also have the opportunity to travel to a foreign country, live with a host family, and attend a year of high school studying an agricultural curriculum.

There are several ways for an FFA member to become involved with international activities. *Courtesy of Carol Duval, National FFA Organization.*

Other ways to become involved with international agriculture are through study seminars and work experience programs. Agricultural study seminars are organized annually for state associations. The seminars last two to three weeks and usually include much traveling. The seminars typically include visits and tours through agribusiness industries, research labs, agriculture schools, and a wide diversity of farms. Work Experience Abroad (WEA) is an international experience program for FFA members in college or alumni FFA members who work in the agriculture industry. FFA members are placed in other nations to live and work on farms or agribusiness for three to twelve months. The program also coordinates the hosting of students from other nations who will come to the United States. The experience of hosting an exchange student can be as meaningful as traveling abroad.

from the cattle industry (Figure 1–9). The swine industry gives us such meat products as pork chops, ham, and sausage. The sheep industry supplies meat for a large part of our population. The poultry industry produces meat for such products as fried chicken and chicken pot pies. The poultry industry

FIGURE 1–9
Steak for our dinners comes from the cattle industry. *Courtesy of Progressive Farmer.*

provides a source of nutritious, inexpensive meat and protein-rich eggs. Fish, shellfish, and other aquatic animals are grown by producers to be used as food. The dairy industry supplies us with fresh milk, cheese, butter, cream, yogurt, ice cream, and other products.

The animals that are slaughtered give us other products. Leather for making shoes, belts, and other clothing comes from cattle, hogs, and other animals. In addition, several medicines and other by-products are made from animal products.

Animals are still used for work. In many places in the western part of our country, horses are used to herd cattle. They are used as pack animals. Horses provide us with recreation through riding and horse racing (Figure 1–10). Dogs are also produced for work. They are valuable in herding sheep and are used in moving cattle and hogs.

An increasing part of the animal industry is the raising of companion animals. Dogs are raised and trained to be used to help visually impaired people, and are raised for pets. The pet industry is broad and diverse. Animals from goldfish to cats are raised for companions to humans.

FIGURE 1–10
Horses provide us with recreation. *Courtesy of Michael Dzaman.*

Like the plant industry, the animal industry has a lot of support areas. Animals have to be fed. The feed industry supplies producers with high-quality feeds that contain the nutrients animals need (Figure 1–11). Medicines are needed to keep the animals healthy. Pesticides are needed to keep the animals free from parasites. Fencing has to be produced. Buildings and facilities have to be built to house the animals. Trailers and trucks are needed to transport the animals. Buyers, brokers, inspectors, and processors all involved in the industry. And it is all a part of agriculture.

Agricultural Exports and Imports

An **export** is a product that is shipped from the United States to a foreign country. An **import** is a product that is brought into this country from another country. Agriculture in this country both exports and imports **commodities**. Commodities are any useful things that can be produced, sold, or bought. Commodities are the end result of agricultural production (Figure 1–12). The trade balance of agricultural products is good news for our country.

FIGURE 1–11
The feed industry provides feed for livestock. *Courtesy of Getty Images.*

FIGURE 1–12
Agriculture in this country imports and exports goods. *Courtesy of James Strawser, The University of Georgia.*

The trade balance refers to the value of goods shipped out of this country (exports) compared to the value of goods shipped into this country (imports). In recent years, some industries have lost a large share of their market to companies in other countries. However, American agriculture still dominates the world's agricultural demand. Our wheat, corn, soybeans, and other products are shipped all over the world to feed people. Each day we export around $6 million in agricultural products to other countries around the world. This generates more than $100 billion annually for our economy by creating business activities and millions of jobs.

Some agricultural products are brought into this country. Although we have a near ideal climate for agricultural production, some crops grow better elsewhere. For example, coffee grows best in the mountainous areas of South America (Figure 1–13). Bananas grow in the tropical areas of Central America, and many specialty foods such as cocoa and vanilla are brought in from other countries.

Another reason for importing agricultural products is so that we can enjoy fresh produce year-round. Because much of South America is south of the equator, their seasons are opposite from ours.

FIGURE 1–13
Coffee is a crop that does not grow well in the United States. *Courtesy of James Strawser, The University of Georgia.*

This means that when it is winter here it is summer there. We take advantage of this by bringing such produce as grapes into the United States so we can enjoy them all year round. Countries such as Mexico that are closer to the equator have a longer **growing season** and can produce vegetables and melons earlier than producers in the United States. By shipping them in from Mexico, we can enjoy watermelons even before it is time to plant them in most parts of this country.

All of the industries involved in exporting and importing agricultural products create a lot of jobs. This adds to the economy and well-being of our nation. When the imports of manufactured products exceed the exports, agricultural exports help to make up the gap.

Distribution

If you are an average person you probably eat beef from Iowa, apples from Washington State, cheese from Wisconsin, tomatoes from California, potatoes from Idaho, and oranges from Florida. You wear clothes made of cotton from Arizona and live in a house built from lumber grown in Oregon. Of course, other states also produce these products—agricultural products come from a wide area. The states mentioned are ideal for the production of these items because of their climate and other factors. Cotton grows poorly in Washington State, and Arizona is more suited to growing cotton than timber.

Agricultural products are grown where the climate and other factors are suitable for production. However, people in Arizona need lumber, and people in Washington State need cotton. To solve this problem, agriculture has a marvelous system for distributing products all over the country (Figure 1–14). Many people make their living buying farm produce

FIGURE 1–14
Agricultural products are shipped from one part of the country to another. *Courtesy of James Strawser, The University of Georgia.*

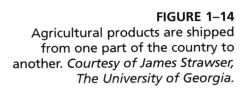

FIGURE 1–15
Americans prefer to buy their food ready or almost ready to eat. *Courtesy of James Strawser, The University of Georgia.*

from producers and selling to wholesale distributors or to processors. These people process and package the produce and sell to retail stores that offer the produce to the public. All this involves a complicated network of transportation. The produce is carried by railroad or by truck. Produce that is easily spoiled may be transported in a refrigerated railcar or truck.

Food and fiber processing is a huge industry that is a part of agriculture. Consumers want products that are as nearly ready to eat as possible (Figure 1–15). People no longer want to wash, shell, and cook beans in the hulls as they come from the producer. Instead, they prefer to have the beans processed and frozen in packages that go into the microwave oven and come out ready to eat. Few people want to make a pizza from scratch when they can have one that is ready to be popped into the oven for a few minutes. Think a minute about all of the processing that had to go into making that pizza. The tomato sauce had to be made from raw tomatoes and spices. The crust was made from flour that was processed from wheat. The cheese had to be processed from milk, and the sausages had to be processed from various meats and spices.

All of the processing required that large numbers of people work together just to get the pizza ready for you—and they are all a part of agriculture.

Summary

Look around you. Agriculture is very much a part of your life. Without it our lives would be dramatically changed. In the following chapters we will explore the wonderful world of American agriculture.

CHAPTER REVIEW

Student Learning Activities

1. Make a list of all the jobs in your community that are involved in agriculture. Choose a particular job and list all the qualifications one would need for the job.

2. Go to the library and research the agricultural products grown in your state. Write an essay on these products. Which are the most important? Are the products used within the state, or are they shipped to other places?

3. Interview 10 people about their perceptions of agriculture. Draw a conclusion as to whether or not the majority of the perceptions are correct.

4. Keep a journal of all your activities for a week. At the end of the week, list all the ways agriculture has affected your life.

True/False

1. ___ Physics, chemistry, geology, meteorology, and biology are all part of agriculture.

2. ___ The foundation of all agriculture is growing plants.

③ ___ Most of the animals produced in this country are raised for producing dairy products.

④ ___ U.S. agricultural imports greatly exceed exports.

⑤ ___ Compared to people in other countries, we pay a very large portion of our income for food.

⑥ ___ Soil and climate provide the basis for the agricultural industry.

⑦ ___ A large portion of the medicines we use comes from plants.

⑧ ___ No other country produces more agricultural products than the United States.

⑨ ___ Companion animals assist farmers with work on the farm.

⑩ ___ One large and expanding aspect of agriculture is the growing of ornamental plants.

Multiple Choice

① The study of or the explanation of natural phenomena is called
   a. agriculture
   b. science
   c. farming

② One American farmer feeds more than
   a. 100 people
   b. 200 people
   c. 500 people

③ The country that leads in the production of cotton is
   a. the United States
   b. India
   c. China

④ Most of the animals produced in this country are raised for
   a. work
   b. dairy products
   c. food

⑤ No other discipline deals more directly with the laws of nature than does
  a. agriculture
  b. horticulture
  c. floriculture

⑥ Agriculture is the most important of all the sciences because we depend on it for
  a. dairy products
  b. jobs
  c. food

⑦ The foundation of all agriculture is
  a. raising animals
  b. growing plants
  c. growing fruit

⑧ The seeds of plants supply us with a rich source of
  a. carbohydrates
  b. protein
  c. vitamins

⑨ Each year many people make their living raising plants in
  a. greenhouses
  b. their homes
  c. gardens

⑩ Pesticides are needed to keep animals free from
  a. predators
  b. parasites
  c. poor weather conditions

Discussion

① How are medicine and ecology related to agriculture?

② List some jobs involved with agriculture.

③ Why are some plants so important in our daily lives?

④ What are companion animals?

⑤ What is needed to raise healthy animals?

⑥ List the uses we have for animals.

⑦ Why is agriculture the most important of all sciences?

⑧ Where does chewing gum come from?

⑨ Why is food and fiber processing such a large part of agriculture?

⑩ List the three applications available for all of our knowledge about living things.

CHAPTER 2

The History of Agriculture

Student Objectives

When you have finished studying this chapter, you should be able to:

- Describe the importance of agriculture.
- Discuss how agriculture was necessary to the development of civilizations. # 13
- Tell how the cotton industry brought wealth and created problems in the United States. # 14
- Describe how the cattle industry developed.
- List some of the inventions that had an impact on agriculture.
- Discuss how scientific research helped develop agriculture.
- Define the Land Grant concept of education.
- Explain how the Land Grant institutions have helped shape agriculture.

Key Terms

civilizations	scientific research	USDA
hunters and gatherers	Land Grant Act	experiment stations

The science of agriculture is almost as old as the human race. In the history of humans on the planet, many **civilizations** have come and gone. Some were powerful and warlike and some were small and peaceful. Some civilizations, such as the Greeks and Romans, provided the foundation for future civilizations. Architecture, art, literature, and political theories came from these civilizations. Other peoples accomplished great feats of engineering and building. Many aspects of ancient civilizations were different, but they all had one thing in common. Before any civilization could exist and flourish, a strong agricultural base had to be established (Figure 2–1). The reason for this was simple: Before roads could be built, buildings designed or works of art created, the people had to be fed.

Before the creation of agriculture, people had to spend most of their time hunting and gathering food. These people are known as **hunters and gatherers**. In order to survive, they had to find and collect the food nature had to offer. This was a full-time job for all people, and little time could be spent doing anything else. This caused the people to have to travel all the time because food in an area would soon be depleted. When people found they could

FIGURE 2–1
Before any civilization can flourish it must have a strong agriculture.
Courtesy of Darl Snyder, The University of Georgia.

FIGURE 2–2
When people could grow their own food, they could stop wandering and build villages. *Courtesy of Darl Snyder, The University of Georgia.*

grow their own food, many of their problems were solved. Now they have a ready supply of food, and could now settle down in one place (Figure 2–2).

Shortly after people began to plant seed and tame animals, they began to search for better ways of growing food. By trial and error they discovered the best time to plant, the best plants to grow, and the best animals to raise. As they got better at growing plants and raising animals, more food could be produced. As more and more food was produced, it took fewer people to grow the crops and keep the animals. In time, only certain people grew food, and others were then free to accomplish other things. A system of trade was established where goods such as pottery or clothing were traded for the food produced by the farmers. When farmers became really proficient, people could settle in one place and create villages. They could then build roads and towns and create art. This was because other people produced the food needed to feed the workers and builders. The stronger the agriculture of a civilization, the stronger could be the army and groups of workers. Without plenty of food to feed all of the people, little progress can be made by any people.

(handwritten margin notes)

Set crop supply

Demand USA   Demand England

more people + more people

S   N   Demand Increases

Need Technology/Development to ↑ Supply to meet the

↑'d Demand. to make

Supply = Demand

American Agriculture

The United States is among the wealthiest nations on the earth. Much of this wealth can be attributed to the tremendously successful agriculture system. As pointed out in the first chapter, American agriculture is the envy of the world (Figure 2–3). No one even comes close to matching the agricultural production of this country.

Our agricultural system began before we were a country. Most of the immigrants who first came to the New World were farmers. The colonists made their living producing crops that could be sent back to Europe. At the time of the American Revolution, well over 90 percent of the colonists made their living through agriculture. In fact, most of the signers of the Declaration of Independence were farmers. At this time the main crops grown were tobacco, sugarcane, and rice. The produce that was not used by the colonists was sent to England, where the products were exchanged for manufactured goods. The rich soil and climate made the New World an ideal place for agriculture. These conditions allowed the development of a variety of crops. But before agriculture could advance very far, certain developments had to take place. The following is a description of some of these developments.

FIGURE 2–3
American agriculture is the envy of the world. *Courtesy of John Deere Company.*

★ King Cotton ✦ # 14

Perhaps the first huge impact that agriculture had on the wealth of the nation came about as a result of cotton production. Cotton had been grown in India, Egypt, and China for thousands of years. It had also been grown in the New World in what is now known as Peru and Mexico from 2500 B.C. It was introduced into the coastal areas of Georgia and South Carolina in very early colonial days. The fiber from the plant was much prized as a material to use in making clothes. It was easy to spin and weave, easy to dye, and comfortable to wear.

There were two types of cotton—upland and Sea Island cotton. Sea Island cotton was the most widely grown. The fibers were longer (called stable length), and the seeds could be easily removed from the fibers. The problem with upland cotton was that the fibers seemed to be "glued" to the seeds and were extremely difficult to remove (Figure 2–4). Even though it was a slow process, the seeds from the Sea Island cotton separated from the seeds by hand very easily. The problem was that this type cotton would only grow along the coasts of the southern colonies.

About 1790, an inventor named Eli Whitney built a machine called a cotton engine (the name was later shortened to cotton gin) that could remove the seeds from upland cotton (Figure 2–5). The device consisted of spikes on a revolving drum that passed through slots and separated the lint from the seeds. A rotating brush then removed the lint from the seeds. One person turning the crank of the gin could remove more seeds than many dozen people removing them by hand. When the gin was fully developed and powered by steam, removing seeds was no longer an obstacle to growing cotton. This new device had a profound effect on the history of our country. For the first time, upland cotton could be profitably grown. This meant that a large area

FIGURE 2–4
Upland cotton has seeds that seem to be glued to the lint. *Courtesy of James Strawser, The University of Georgia.*

FIGURE 2–5
The cotton gin removed seeds from the cotton lint. *Courtesy of National Cotton Council of America.*

FIGURE 2–6
One family could clear the land and make a living producing cotton. *Courtesy of National Cotton Council of America.*

from what is now southern Missouri to Virginia and all of the areas south of there could be used to produce cotton. Cotton was an ideal crop to grow. Only so much tobacco, sugar, and rice could be used and the market could be easily flooded. But cotton was a product that everyone could use. Now that the seeds could be efficiently removed, a gigantic market opened for cotton. As the demand grew, more cotton was produced. When this happened, the price fell. This caused the cloth to be affordable to poor people as well as rich. This type clothing was cheaper than the traditional wool or linen. It also had characteristics such as comfort that people wanted. When the price fell, more people could buy it and demand increased.

As more and more cotton was produced, new fields to the west in Alabama, Tennessee, and Mississippi were cleared and planted. A family could clear the ground and plant, cultivate, and harvest enough cotton to make a living (Figure 2–6). Although a lot of cotton was produced by large plantations using slave labor, even more cotton was produced

FIGURE 2–7 In the mid-1800s most of the cotton produced in the United States went to Europe. *Courtesy of National Cotton Council of America.*

by families who worked the land themselves. The entire southern portion of the United States was settled primarily because of the cotton industry. Cotton became so important that it became known as King Cotton.

Most of the cotton produced went to Europe (Figure 2–7). The textile factories of England, France, and Germany ran night and day using cotton from the United States. Ships that delivered the cotton returned with manufactured goods that were sold to the people of the southern states. By the 1850s these industries were almost totally dependent on American cotton.

While the South was developing its economy on cotton, the northern states were developing industries. These two diverse sections of the country developed disputes over the role of the federal government, taxation, and about the issue of slavery. Ultimately, these differences led to the American Civil War. The development of agriculture had had a profound effect on the young country!

The Cattle Industry

The cattle industry in this country had a slow beginning. Beef was expensive. In the areas close to population centers, cattle had to be fed expensive grain or had to be fed on pastures that could be put to better use growing crops. In the newly settled areas of the West, cattle could be raised on vast areas of grass (Figure 2–8). These areas were suited to the growing of cattle but not crops. The only problem was that the cattle-grazing areas were far from the centers of population.

After the American Civil War, western cattle producers decided the best way to get their cattle to the large cities was to ship cattle on the new railroads that had recently stretched westward. This would provide a way of getting the live cattle close enough to the consumer (Figure 2–9). This was necessary because the animals had to be slaughtered and consumed within a short period or the meat would spoil. The big problem was that there were only a few places where the railroad ran. Certain areas such as central and western Texas had no rails at all. Towns like Sedalia, Missouri, and Abilene,

FIGURE 2–8
In the newly settled territories of the West, cattle could graze on the vast areas of grass. *Courtesy of Texas and Southwestern Cattle Raisers Foundation, Ft. Worth, TX.*

FIGURE 2–9
Cattle were driven to railheads where they were shipped to the East. *Courtesy of Texas and Southwestern Cattle Raisers Foundation, Ft. Worth, TX.*

Kansas, became established as "railheads" where cattle were brought for shipment. The only way to get the cattle to the railheads was to drive them. The fabled cattle drives of the West were started in this manner. For a period of about 20 years, this was the means of getting cattle to market. This allowed cattle producers to make a profit by using the millions of acres of grassland to graze cattle.

After the railroads were finally completed around the turn of the twentieth century, cattle could be delivered live without the long arduous drives. A problem still remained. Even the rail journey took several days. During this time, the cattle lost weight from stress of the long trip. Producers reasoned that more profit could be made if the cattle were slaughtered close to where they were raised and the meat shipped. However, this was impossible because the meat would spoil before it could reach market.

People had known for centuries that if meat was kept cool it could be kept longer without spoiling. The first attempt at cooling meat was the use of ice that had been cut from frozen lakes during the winter and stored in icehouses. The ice blocks were suspended from the ceiling in the meat storage rooms in an effort to keep the meat cool. This effort was not very successful. The ice melted quickly and did not keep the meat cool enough. Then, during the 1880s, mechanical refrigeration was developed. This allowed storage rooms to be kept cool all year long.

A few years later the refrigerated box car was invented. This innovation allowed the transportation

Career Development Events

The National FFA is an organization for students studying agricultural education in high school. This organization is active in all 50 states and Puerto Rico. Over 450,000 students participate in activities and programs sponsored by the FFA.

This organization began in 1928 when students studying agriculture met in Kansas City for the purpose of establishing a national association. Prior to this, many states had created similar organizations for rural young people. The state organizations began to compete in livestock judging and looked for a place to hold a national competition. The American Royal in Kansas City agreed to host the event. This brought the students together for the first time, and they decided there was a need for a national organization. In 1928 the Future Farmers of America came into existence.

The FFA was patterned after the Future Farmers of Virginia. This state association was developed by Henry Groseclose, Walter Newman, Edmund Magill, and Harry Sanders. They saw the need to give rural youth opportunities to develop social graces and leadership skills. Ceremonies for the FFA were designed after the ceremonies of the Freemasons, and the awards were patterned after the Boy Scouts.

The FFA has passed many milestones since 1928. In 1965 the FFA merged with the New Farmers of America (NFA). The NFA was an organization for black youths studying agriculture. In 1969 girls were allowed to join the organization. In 1988 the name was changed from the Future Farmers of America to the National FFA Organization.

From 1928 until 1998, the FFA held its annual convention in Kansas City. Beginning in 1999, the convention was moved to Louisville, Kentucky. Around 50,000 young people spend several days in competition, leadership seminars, award programs, and instructional and motivational programs. This is the

of meat anywhere in the country anytime during the year. Now not only could animals be slaughtered any time of the year, but also the meat could be stored for a long period of time. This meant that meat could be distributed to everyone in the country. This had the added effect of lowering the price of meat. Americans began to enjoy a healthy diet at a relatively low cost.

world's largest youth convention. The FFA is constantly updating and expanding programs to better serve students. Whether you wish to enter individual competition, team competition, work toward the several degrees of membership, or travel to countries all across the world, there is a program for your needs. There is a place in FFA for you!

The FFA began in 1928 and was organized for students studying agriculture. *Courtesy of the Agricultural Education Program, The University of Georgia.*

Crop Production

Advances also were made in the producing of crops. As people moved westward, vast areas of fertile land were opened for farming. A particularly fertile area was the grassy plains of the Midwest. When settlers first came into the area, they had difficulty plowing the soil. The heavy soil could not be turned with the wooden plows used by the settlers. A heavy cast iron plow would not work because animals could not pull it through the sod (Figure 2–10). The problem was solved in the 1830s by a man named John Deere. He developed a plow with a cutting edge and share made from steel like that used for saw blades. The strong tempered steel allowed the plow to be thinner and lighter (Figure 2–11). Wheels were later mounted to make the plow easier to pull by animals. With this invention, the fertile ground could be plowed under and cultivated.

Another problem was that of harvesting crops. For example, wheat was difficult to harvest because the plants had to be cut by hand. They were tied

FIGURE 2–10
Settlers had difficulty plowing the thick sod of the Midwest. *Courtesy of Dr. David C. Hartnet.*

FIGURE 2–11
The tempered steel of Deere's plow allowed it to be lighter. *Courtesy of John Deere Company.*

FIGURE 2–12
McCormick's reaper made harvesting grain more efficient. *Courtesy of Case-International Harvester.*

together into bundles called shocks and stacked. They were then gathered and threshed until the grain left the plants. This took a lot of time and reduced the amount of wheat that could be grown.

In 1831, Cyrus McCormick invented a machine that was used for reaping wheat (Figure 2–12). The horse-drawn implement cut wheat that previously had to be cut by hand. It operated by means of a sickle cutter that was powered by the turning of the wheels. Soon threshing machines were developed to separate the grain from the rest of the plant. This machine was run by steam power and people fed wheat shocks into the machine and it threshed the wheat grain out. After the invention of the internal combustion engine, the process of cutting the wheat and threshing it was generally combined into one operation with a machine that came to be known as a combine (Figure 2–13).

The development of the internal combustion engine had a revolutionary effect on agriculture. After its development, work that once took days using human or animal power could now be done in a matter of minutes and with a much better result. Today almost all operations involved in the production of agricultural products are mechanized.

FIGURE 2–13
A modern combine cuts and threshes the grain. *Courtesy of John Deere Company.*

Scientific Research

All of the developments mentioned so far had a real impact on American agriculture. However, by far the most important aspect of the development was scientific research. **Scientific research** is the systematic search for new knowledge. As mentioned earlier, better ways of growing plants and animals were developed by trial and error. However, this method was too costly and took too long. Dramatic progress was not made until the development of methods of scientific research.

Systematic scientific research began in this country about the middle of the 1800s. It had become apparent that agriculture was a major economic basis for the nation. Leaders thought that our abilities should be strengthened, and the best way to do that was to educate people about agriculture. At that time, the universities in the United States taught a curriculum known as the classics. Students studied subjects such as Latin, Greek, history, philosophy, and mathematics. Though these were challenging disciplines, they had little practical use. People began to realize there was a need for institutions of higher learning where students could study subjects such as agriculture that had a practical application.

In the late 1850s a senator from Vermont, Justin Morrill, introduced a bill to provide public land and funds to establish universities to teach practical methods of producing food and fiber. The bill passed in 1862 and became known as The **Land Grant Act** or the Morrill Act. This act provided each state with public land to build a "college for the common people" (Figure 2–14). The purpose of these colleges was to teach agriculture and mechanics. Many of these institutions were called "A&M" universities. Many, such as Texas A&M, still use this name. During that same year, President Lincoln signed into law a bill that established the United States Department of Agriculture **(USDA)**. Soon almost all of the states in the country established Land Grant colleges.

As students enrolled and classes began, a severe problem was recognized. People began to realize they had little knowledge about agriculture. Most of the knowledge about growing plants and animals had been obtained through trial and error.

Pg. 33

FIGURE 2–14
In 1862 Land Grant universities were established. *Courtesy of Agricultural Education Department, The University of Georgia.*

FIGURE 2–15
Experiment stations provided research on new and better ways of production. *Courtesy of Stan Diffie, The University of Georgia.*

Most of the ideas about agricultural production had no real scientific basis. To solve this problem, Congress passed the Hatch Act in 1872, which authorized the establishment of **experiment stations** in states with land grant colleges. The purpose was to create new knowledge through a systematic process of scientific investigation (Figure 2–15). These experiment stations put to use what has come to be known as the scientific method of investigation. Here a systematic, scientific investigation about how plants and animals grow was conducted. Most of the progress of American agriculture came about as a result of the research at Land Grant universities and their experimental stations.

In 1890, another Land Grant Act was passed that provided additional colleges. These served black people who lived in the South. These colleges and universities have made a tremendous contribution to the research and teaching of agriculture. Today, the "1890" Land Grant colleges are open to all people.

In 1914, Congress passed the Smith Lever Act, which set up the Cooperative Extension Service. This legislation placed an extension agent in each county in each state. The purpose was to teach the new information learned from research to people. This completed a system known as the Land Grant concept. This concept was that the purpose of a Land Grant university was to teach, conduct research, and carry the new information to the people in the state through the Extension Service.

In 1917, the Smith Hughes Act established Vocational Agriculture in the public high schools as a means of teaching new methods of agriculture. This act established the class you are now in.

Changes in Agriculture

In the past, agriculture has seen many changes. Some of these changes have taken place slowly, whereas others have happened relatively quickly. Modern agriculture is changing at the most rapid pace in history. In fact, you will probably see more change in the way we grow food and fiber than any generation in history.

Scientists are on the verge of being able to locate each of many specific genes on plant and animal chromosomes. In fact, the entire gene sequence of several plants and animals have already been completely mapped. Hundreds of different genetically engineered crops are currently being grown. When scientists are able to map all of the genes on chromosomes, the potential is almost unlimited. Literally thousands of different combinations of gene splices can be accomplished that will make plants and animals more efficient to grow. In addition, the nutritional value of our food supply will be enhanced. Chapter 20 will explain more about these advancements.

Food products will also change. The trend over the past few years has been toward the sale of processed or semiprocessed food. Consumers want food that is quick and easy to prepare. In the future we will see less and less fresh meat and vegetables sold in the grocery store. Food that can be placed in the microwave for a few minutes and come out as a ready-to-eat dinner will be the way most food is prepared.

There will likely be more changes. Most of the changes we will see during the next 20 years, we can't even imagine. Scientists will find newer and better ways to grow, process, and distribute food and fiber.

Summary

The development of new methods through research and the teaching of this knowledge helped bring about progress in agriculture. Developments such as pesticides, new and better varieties of plants, fertilizers, and soil conservation all resulted from scientific research. As you study the remaining chapters, keep in mind that the agricultural industry is the backbone of our nation. This powerful industry was built as a result of scientific research and development.

CHAPTER REVIEW

Student Learning Activities

1. Interview several elderly people who have lived in your community for many years. Ask them about the importance of agriculture to the development of the community. Report your findings to the class.

2. Pick a particular segment of agriculture and do further research on how that segment developed. Discuss your findings with the class.

3. Contact the Cooperative Extension Agent in your area. Discuss some of the latest agricultural research being done or that is recently completed at the Land Grant university in your state. Discuss the impact this research might have on agriculture in your area. Report to the class.

4. Research an ancient civilization and report on the role agriculture had on the development of the society. What crops and animals did they grow? What techniques did they use to produce them?

True/False

1. ___ Before any civilization could exist and flourish, a strong agricultural base had to be established.

2. ___ At the time of the American Revolution, 50 percent of the colonists made their living through agriculture.

③ ___ The main crops during the American Revolution were tobacco, sugar-cane, and rice.

④ ___ The rich soil and climate made the New World an ideal place for agriculture.

⑤ ___ Sea Island cotton fibers were extremely difficult to separate from the seeds.

⑥ ___ Eli Whitney developed a machine to remove seeds from cotton.

⑦ ___ The northern states developed their economy around industry.

⑧ ___ President Lincoln signed into law a bill that established the United States Department of Agriculture.

⑨ ___ Most of the immigrants who first came to the New World were industrialists.

⑩ ___ Most of the cotton produced went to Europe.

⑪ ___ Food will probably change very little in the coming years.

⑫ ___ Genes can be located on the chromosomes by mapping.

Multiple Choice

① The civilizations that provided a foundation for future civilizations were the Greeks and the
    a. Romans
    b. Americans
    c. Europeans

② Crops were sent to England and exchanged for
    a. wheat reapers
    b. cotton gins
    c. manufactured goods

③ In colonial times, the most widely grown cotton was
    a. King Cotton
    b. manufactured cotton
    c. Sea Island cotton

④ The invention that allowed the transportation of meat anywhere in the country, anytime during the year, was the refrigerated
    a. cotton gin
    b. box car
    c. "railhead"

⑤ A machine was developed for reaping wheat by
    a. Cyrus McCormick
    b. Eli Whitney
    c. John Deere

⑥ The most important aspect in the development of agriculture was
    a. wheat reaping
    b. scientific research
    c. colony industrialization

⑦ Another name for the Land Grant Act was the
    a. Morrill Act
    b. Hatch Act
    c. Hughes Act

⑧ The Smith Hughes Act established
    a. manufactured goods
    b. scientific research
    c. vocational agriculture

⑨ The biggest problem cattle producers faced was
    a. research
    b. transportation
    c. climate

⑩ The Smith Lever Act set up the
    a. Cooperative Extension Service
    b. refrigerated box car
    c. Smith Hughes Act

⑪ Scientific Agricultural research began in this country during
    a. the middle 1700s
    b. the middle 1800s
    c. the middle 1900s

⑫ The cotton gin
   a. picked cotton.
   b. made cloth from cotton fibers.
   c. separated fibers and seed.

Discussion

❶ What problem did cattle producers have?

❷ Why is John Deere important?

❸ What is the Land Grant Act?

❹ What is the Land Grant Concept?

❺ Why was cotton so important in the South?

❻ Why is agriculture important?

❼ Why was agriculture necessary to the development of civilizations?   Pg. 20

❽ How does the cotton gin operate?

❾ How did scientific research help develop agriculture?

❿ Why did cotton become so important?

CHAPTER 3

Soil: The Origin of Life

Student Objectives

When you have finished studying this chapter, you should be able to:

- Explain why agriculture and all of life depend on the soil.
- Explain the difference between organic and inorganic soil.
- Discuss the carbon cycle.
- Discuss the different ways soils are formed.
- Contrast the differences between acidity and alkalinity and the effect of each on the soil.
- Explain the concept of a soil ecosystem.
- Conceptualize the balance of organisms in the soil.

Key Terms

soil	minerals	texture
food chain	glaciers	soil horizon
humus	erosion	ecosystem
organic matter	flood plain	

The earth has a thin coating of material that is called **soil**. Imagine that the earth is an apple. The layer of soil would be as thin as the peel of the apple. Yet without this layer, there could be little life on the earth. Most plant life is dependent on the soil. Plants are rooted in the soil and get their life-sustaining nutrients from it. If there were no soil, there would be no plants for land animals to eat. Life is in a constant cycle with the earth. A process called the **food chain** begins with the soil. A plant obtains nutrients from the soil and uses them through energy from the sun. To obtain the nutrients it needs, a mouse may eat the plants. An eagle then eats the mouse and obtains nourishment. Eventually the eagle dies and its body is decomposed. The decaying eagle returns nutrients to the soil and the chain begins again. In this process all life is dependent on the soil (Figure 3–1).

Soil is composed of four ingredients: minerals, air, water, and **humus**. Humus is another word for **organic matter**, and **minerals** are substances that did not come from living organisms. Minerals have the chemical and physical properties that can be formed into rocks. Minerals in the soil are very small particles. As these fine particles are grouped together, they form pores or gaps where the particles touch. These pores are filled with either air or water.

Organic Soils

Organic matter comes from living sources. In the example of the eagle that dies, the material left when it decays is organic material or humus. Soil can be grouped into two very broad categories based on how much mineral or humus the soil contains. Soil that has a lot of humus or organic material is called organic soil. This soil is the result of a buildup of plant materials that grew in the soil,

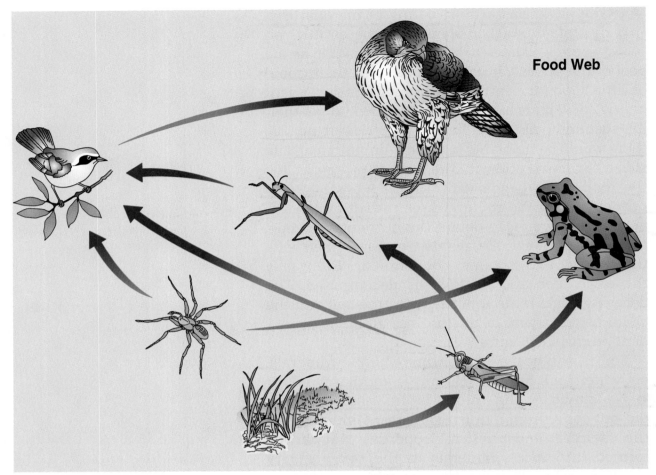

Food Web

FIGURE 3–1 In the food chain, nutrients are passed from the soil to living creatures and back to the soil.

died, and decayed (Figure 3–2). Over many hundreds of years, the decayed plants turned into a black fertile soil called organic soil. Organic soils exist in parts of the Midwest where centuries of grasses grew and returned to the soil when they decomposed. Another area with these soils is southern Florida, where swamp land has been drained. For thousands of years, a type of grass called saw grass grew in the water. When it matured, it fell into the water and decayed. Over time, several feet of this organic material turned to soil.

FIGURE 3–2
Organic soil was formed by hundreds of years of decaying grasses. *Courtesy of Getty Images.*

Organic soil is very rich in nutrients. Also, the organic material in the soil allows nutrients in fertilizer to be available for plant use. Organic soils are some of the world's most productive soils.

Inorganic Soils

Soils that originated from minerals are the result of the wearing down or breaking down of rock. This process is often the result of weather. As wind blows against a rock, particles are loosened and removed (Figure 3–3). Likewise rain, sleet, and snow on rocks can remove particles that become soil. The process of wetting and drying as well as the dissolving of minerals by the water can also help in the process of breaking down the stone. The freezing and thawing of water in cracks and crevices of rocks can wear away tiny particles and can open other cracks for the entrance of more water. The process of weathering of rocks is an extremely slow process that occurs over thousands and even millions of years.

Ice may form soil by moving across the surface of the soil. Thousands of years ago, great rivers of slowly moving ice called **glaciers** extended across the northern portions of America, Europe, and Asia.

FIGURE 3–3
Wind is a powerful force in weathering rocks and the formation of soil. *Courtesy of Getty Images.*

As these huge masses of ice crept southward, they carried along rocks that were in the path. As the glaciers moved along, the rocks were crushed together with enough force that some were pulverized and made into particles small enough to become soils (Figure 3–4). With the change of the seasons, part of the ice would melt and then freeze again. During warm periods, the front of the ice mass would melt and drop all the debris it had picked up along the way. When the weather once again turned cold, the ice resumed its movement

FIGURE 3–4
Slow moving glaciers crush and pulverize rocks into soil particles. *Courtesy of W. S. Keller, National Park Service.*

and picked up other rocks and soil particles. This mass of large boulders, smaller rocks, sand, silt, and clay is called till. The land mass built up by the dropping of till by a glacier is called a moraine. Obviously, the type of debris picked up by the glacier determined the type of soil that was deposited. Moraines vary in the type and texture of the soil. Some contain a lot of sand and are coarser, whereas some contain the finer silt.

Water-Deposited Soil

As mentioned earlier, moving water wears away rock and forms soil particles. Soil that has been transported and deposited by moving water is called alluvial soil. Some of the world's most productive soils are the result of alluvial action. Water is a major way that soil is moved from one place and deposited in another. Water in a stream picks up soil from the ground. When rains wash soil particles into the stream, the currents carry the material down the stream (Figure 3–5). This process is called **erosion**. As long as the water is moving rapidly, the soil particles remain suspended in the water. As soon as a streambed begins to level out and the water slows down, the soil particles settle to the bottom. If the stream flows out of its bank, deposits are left in the area covered by the water. These areas are called **flood plains**. Also, as the stream nears its mouth, the water slows down and the deposits settle to the bottom. These areas are called deltas. Gradually over the years, the soil is built up to a great depth by both of these processes.

In this country, the Mississippi River has carried deposits south for thousands of years and deposited them at the mouth of the river. The result has been the rich farmland of the Mississippi Delta (Figure 3–6). Since the early history of our country,

FIGURE 3–5
Water that moves in a stream often picks up and transports soil.
Courtesy of USDA Natural Resource Conservation Service.

FIGURE 3–6
The rich farmland of the Mississippi Delta is the result of soil transported by water. *Courtesy of John Wasniak, Louisiana Agricultural Experiment Station.*

this area has produced rich crops. It should be noted, however, that the soils that made up the Delta were once in the Midwest. The soils in the areas where the soil left are less productive because of the erosion.

Soil may also be deposited into a lake. As a stream empties into a lake, the soil deposits are dropped and they sink to the bottom. Eventually over thousands of years, the lake becomes filled with soil deposits. When the lake disappears, a deep fertile soil is left in place of the water.

Soil Deposited by Wind

The wind also creates and transports soil. In many areas of the world, tall outcroppings of rock have been carved by the wind. As the wind wears away the stone, the loose particles become soil and form deposits known as eolian soils. If the particles are large particles, the soil is called sand and the deposits are called dunes. In parts of the western United States, vast areas are covered in huge sand dunes that are constantly moving and changing shape. Because the sand is constantly moving, little vegetation can grow on the dunes (Figure 3–7).

FIGURE 3–7
Sand dunes constantly move and little vegetation can grow on them. *Courtesy of James Strawser, The University of Georgia.*

Fine soil particles such as silt and clay that have been deposited by the wind create soil known as loess soils. These soils may be found through the Mississippi Valley, parts of the Midwest, Washington, and Idaho. Many of these soils were formed when the glaciers melted and the climatic conditions turned dry. When the soil dried out the fine particles were blown by the wind and accumulated in new areas.

In many of the mountainous regions of the world, soil was (and still is) formed by the action of volcanoes. Volcanoes bring deposits from deep within the earth and spew the material on the surface. Much of this material is in the form of molten rock (called lava). This inorganic material flows out over the surface and cools into solid rock. Other deposits are spewed high into the air. The finer of these particles is called volcanic ash and may be carried many miles by the wind. Near the volcano many feet of volcanic ash may be built up in a matter of days. Areas as far away as a hundred miles may receive measurable deposits of ash. Soils from volcanic ash are generally very productive soils (Figure 3–8). Volcanic soils can be found in the states of Hawaii, Oregon, and Washington.

FIGURE 3–8
Soil from volcanic ash is usually very fertile. *Courtesy of Fred E. Mang, Jr., National Park Service.*

Soil Texture

The **texture** of the soil refers to the size of the individual soil particles. The larger the size of the soil particles, the coarser the soil feels when rubbed between the fingers. The largest of the soil particles is sand. Sand particles are small but they are considered large compared to other types of particles.

Career Development Events

The FFA Land Judging Career Development Event is a competition for teams of four members. Each member competes individually and the final team score consists of the total of the three highest individual scores. The FFA land judging contest aids classroom instruction by teaching the students practical application of the Soil Conservation Service classification system. The competition and hands-on applications motivate students to learn soil and water management practices. A student's decision-making skills are developed and environmental awareness increased through the analysis of soil characteristics and the identification of important soil and water resources.

This event tests the FFA members' ability to determine a site's topsoil texture, thickness, and effective depth of topsoil and subsoil. The student must also decide the permeability of subsoil, land slope, erosion, drainage, and any other factors that will best decide the land's capability. After the student determines land class, the proper land use and treatment practices must be decided. The student judges the land by the observation and measurement of prepared soil pits and the land area surrounding each pit. Each team member should come to the event equipped with a soil probe, pencil, clipboard, and calculator. After all team members have completed each phase of the event, each individual score will be calculated. State winning teams advance to the National Land Judging Career Event.

These activities help make classes more interesting and give students an additional reason for doing their best to develop necessary skills. The Land Judging Career Develepment Event requires students to study classroom material and practice the hands-on activities. The extra time in preparation for the event not only helps produce a winning team but also helps prepare for a career.

The Soil and Water Management Proficiency offers individual FFA

If a soil containing sand is rubbed between your fingers, it will feel coarse and gritty, like corn meal. Silt particles are smaller than sand and feel smoother to the touch. The smallest of the particles is referred to as clay. A soil high in clay content will feel smooth, like flour. One way to illustrate the relative sizes is to imagine that if a clay particle was the size

members an opportunity to be awarded for experiences related to renewable natural resources. Students who have participated in the planning and use management practices to improve air, soil, water, timber, fish, or wildlife can apply for this award. Programs in which

In the Land Judging Career Development Event, students judge the land by the observation and measurement of prepared soil pits and the land surrounding the pits. *Courtesy of Frank Flanders, the Agricultural Education Program, The University of Georgia.*

a member is involved in helping to stimulate public awareness or assisting in educating the public concerning pollution problems should be included in the application. Other activities a student might participate in are the development and implementation of management practices that will prevent erosion, improve the productivity of the soil, promote efficient use of water, or reduce air and water pollution. To compete for the proficiency, the student should keep a list of related skills developed as a result of the above activities and training in vocational agriculture. The student should participate in a Supervised Agricultural Experience Program (SAEP) that relates to soil and water management. Proficiency applications are judged on the student's records, activities, and experiences. Keeping accurate records of accomplishments and soil and water management related income and expenses are essential to becoming a winner.

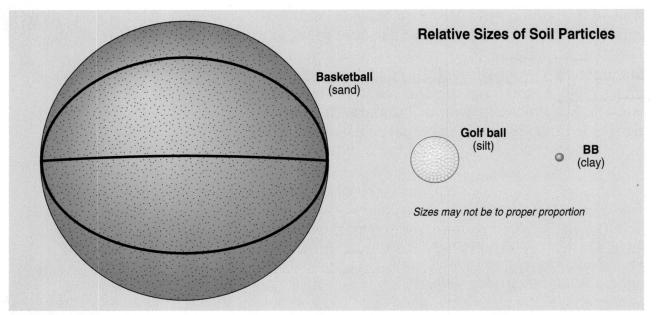

Relative Sizes of Soil Particles

Basketball
(sand)

Golf ball
(silt)

BB
(clay)

Sizes may not be to proper proportion

FIGURE 3–9 If a clay particle were the size of a BB, a silt particle would be the size of a golf ball, and a sand particle would be the size of a basketball.

FIGURE 3–10
Soil with a high clay content does not allow water to pass through very fast. *Courtesy of USDA Natural Resource Conservation Service.*

of a BB, a silt particle would be the size of a golf ball, and a sand particle would be the size of a basketball (Figure 3–9).

Soils are seldom composed of pure sand, pure silt, or pure clay. More often, they are a combination of the three. Soil that contains less than 52 percent sand, 28 to 50 percent silt, and 7 to 27 percent clay is called loam soil. This soil is considered to be near ideal for growing most crops. Soils that have a large concentration of sand are coarse and do not hold water very well. Because the particles are so large, the water from a rain or irrigation system passes on through the soil. Water that goes on through the soil is of little use to plants. Clays and silts slow the water down as it is absorbed into the ground and hold a portion of it that the plants can use (Figure 3–10). If the soil has a large percentage of clay, the ground will not let enough water through. This may either cause the water to run off or may hold the water too long and cause problems for the plant.

Clay also plays an important role in soil fertility. Most producers apply fertilizers to the soil. Fertilizers provide the nutrients needed for the plant to grow. Clay particles in the soil hold the nutrients in place until the plants can use them.

Soil pH

The pH of the soil is a measure of how acidic it is (Figure 3–11). Acids contain hydrogen that forms particles called ions when dissolved in water. Alkaline is the opposite of acid. Vinegar is a common substance that is acidic. Baking soda is alkaline. The pH scale goes from 1 to 14. A score of 7 means that the substance is neutral. The lower the number, the more acid the substance is. The higher the number, the more alkaline the substance.

Certain crops may have problems growing in soils that are too acidic or too alkaline. Producers frequently adjust their soils to the right pH. They have samples of their soil tested for pH. If the soil is too acidic, the correct amount of an alkaline substance (such as lime) is added. If the soil is too alkaline, acid materials (usually sulfur-based materials) are added to the soil. Crops such as blueberries grow well in soils that are relatively acidic (pH of 4.0 to 5.0). Crops such as asparagus do well in soils that are less acidic (pH of 6.0 to 8.0).

Soil Horizons

When soil is formed from rock or from organic material, the soil forms in layers. These layers lie parallel to the surface of the earth. As layers are added, they begin to take on different characteristics. These different regions are called **soil horizons**

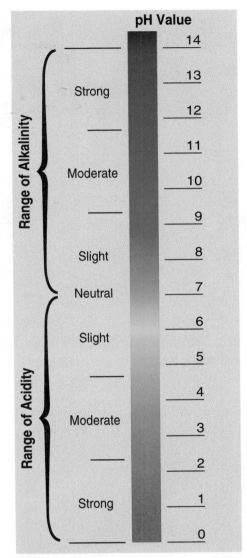

FIGURE 3–11
The pH scale measures the acidity or alkalinity of a substance.

FIGURE 3–12 Soil is classified in soil series. Some examples are: (A) Entisols. (B) Aridisols. (C) Iceptisols. (D) Aflisols. *Courtesy of USDA Natural Resource Conservation Service.*

FIGURE 3–13
A pit dug into the ground can give a view of the soil profile. *Courtesy of James Strawser, The University of Georgia.*

(Figure 3–12). If a pit is dug deeply into the ground, the vertical bank on the side of the hole gives a picture of the different layers or horizons of soil. This picture is called the soil profile (Figure 3–13).

The uppermost layer is called the O horizon. This layer contains undecomposed and decomposed organic matter. This is the place where most seeds germinate.

The next layer is the A horizon and is composed of the top soil that contains organic matter along with minerals. It is generally darker than the horizons below and may be from a few inches to several feet thick.

The region under the A horizon is the E horizon and is the area of the topsoil that has little organic material. This horizon is light in color because the clays and humus have been washed out by water that passes through the soil.

The next region is the B horizon. In this subsoil region, materials washed out (or "leached") from the other horizons accumulate and form. This area is generally high in clay content. The B horizon combined with the O, A, and E horizons make up a region of the soil where plant roots grow.

Underneath the B horizon is the C horizon that is made up of parent material. Parent material is the material from which the soil originated. This horizon has not been affected by the soil-making process and does not have the properties of the soil in the other horizons.

The last horizon in the soil profile is the R horizon. This horizon is composed of the bedrock on which the other soil horizons rest.

The Soil Ecosystem

Your first observation of the soil might lead you to believe that it is lifeless. Looking across a freshly plowed field, you might think that all that is there are minerals and decaying plant matter. Actually, the soil abounds with life. The soil is an **ecosystem**. An ecosystem is all of the plant and animal life that live in an area. The life-forms depend on each other for the proper balance of food and other environmental factors. Within this ecosystem are many different forms of plant, animal, and microbial life. Most of these forms of life depend on each other for their existence or for a maintenance of the proper balance in nature.

Plant Life

Most plants depend on the soil for their existence. The soil provides the support for the root system that anchors the plant and makes it stable. Most of the nutrients used by the plant are obtained from the soil. Even though we are used to seeing the top portions of plants, a very large portion of the plant lives

FIGURE 3–14
A large part of a plant lives in the soil. *Courtesy of USDA Natural Resource Conservation Service.*

in the soil (Figure 3–14). Roots of most plants can reach as far as several feet into the soil. In fact, such common crops as alfalfa have been known to grow as far as 25 feet or more into the ground. The area of the soil that contains the roots of plants is called the rhizosphere. In this zone of the soil the plant receives water and the nutrients it needs to live and grow.

Microorganisms

Within the rhizosphere live billions of microorganisms of many different types. Many of these organisms live off the roots of plants. As plants live and grow, they are constantly oozing materials out through the roots. This material contains protein and other materials that the microorganisms feed on. In addition, as root cells mature and die, the microorganisms act to decompose the cells. When the entire plant dies they help to return the nutrient material the plant took from the soil back to the soil. This is known as the carbon cycle.

The most common soil microorganisms are bacteria. They are so abundant that in only one teaspoon of soil there can be as many as 500 million bacteria. Several types of these bacteria live in what is known as a symbiotic relationship with plants. This means that organisms of different types live together for mutual benefit. The most common symbiotic bacteria are the nitrogen fixing bacteria, rhizobia, that live in the roots of certain plants. Plants that host these organisms are called legumes and include such plants as beans, clovers, peanuts, alfalfa, and peas. Rhizobia live in lumps on the roots called nodules. The bacteria live in the soil and attach themselves to the root hairs of the legume shortly after the plant sprouts. As a reaction to the bacteria attaching itself to the root, a gall or knot forms on the plant root. It is within this swelling that the rhizobia bacteria grow and reproduce. They receive all the nutrients they need to live and reproduce from the host plant. In return the bacteria convert nitrogen from the air in the soil into a form of nitrogen that the plant can use (Figure 3–15).

Fungi are plantlike organisms that contain no chlorophyll. They range in size from microscopic to the large mushroom fungi that grow on the surface of the soil or on decaying plant material. These organisms are as abundant as bacteria in the soil. They play an important role in the breakdown and decay of plant materials (Figure 3–16). Fungi are particularly important in forest soils because they break down lignin, which is a primary component of wood. Trees that die and fall to the ground are returned to the soil through the action of fungi, bacteria, and other organisms.

Protozoa are one-celled organisms that live in moist soil. They are aquatic organisms because they live in particles of water in the soil. When the soil becomes dry, the protozoa change into an inactive state until the soil becomes moist again. Protozoa feed on bacteria in the soil. Through this process, a balance of bacterial life in the soil is better maintained.

One of the most important groups of microscopic animals that inhabit the soil is nematodes. Nematodes are worms in the class Nematoda that have smooth round bodies that are not segmented. Although many nematodes are not microscopic, most of those that live in the soil are so small they can't be seen with the naked eye. These tiny worms are very abundant. In a typical spade full of moist soil there may be over a million nematodes. In fact, they are the most abundant multicelled animals in the soil (Figure 3–17). There are three basic groups of soil nematodes. One group consumes decaying organic matter. Another group eats other microorganisms.

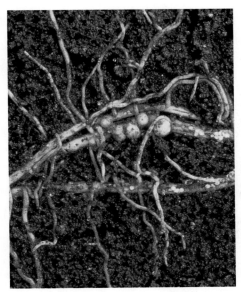

FIGURE 3–15
Bacteria, called rhizobia, form nodules on roots. They convert nitrogen from the air to a form useful to the plant. *Courtesy of USDA ARS.*

FIGURE 3–16
Soil fungi help to decompose plant material. *Courtesy of James Strawser, The University of Georgia.*

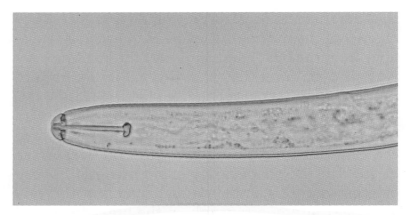

FIGURE 3–17
Nematodes can feed on the roots of plants. They are the most numerous multicelled animals in the soil. *Courtesy of Ed Brown, The University of Georgia.*

The third group feeds on plant roots. The third group is by far the most important because of the damage they do to the plants.

Summary

Soil is essential to all life. All of our food sources depend on it. Humans must protect it because it takes hundreds or even thousands of years to build soil. The fertile soil of this nation has provided much of our wealth. If we take care of our soil, our wealth will continue.

③ CHAPTER REVIEW

Student Learning Activities

❶ Bring in a sample of soil from around your home. Determine if the soil is organic or inorganic. Also determine the soil texture and if the soil is mostly sand, clay, silt, or loam.

❷ Observe the area around your school. Attempt to determine how the soil might have been formed in the area. What indications did you use?

③ Make a list of all the ways soil is important to you. Compare your list to others in the class.

④ Locate soils that are different colors. Determine what makes the difference in color.

True/False

① ___ Most plant life is dependent on the soil.

② ___ If there were no soil, there would be no plants for land animals to eat.

③ ___ A process called the food chain begins with the soil.

④ ___ Minerals have the chemicals and physical properties that can be formed into soil.

⑤ ___ Organic soil is rich in nutrients.

⑥ ___ The largest of the soil products is sand.

⑦ ___ If the soil is too alkaline, acid materials are added to the soil.

⑧ ___ The most common soil microorganisms are the bacteria.

⑨ ___ Soil is made up of only minerals and decaying plant matter.

⑩ ___ Roots of most plants cannot reach very far into the soil.

Multiple Choice

① A plant obtains nutrients from the soil and uses them through energy from
   a. water
   b. food
   c. sunshine

② Humus is another word for
   a. organic matter
   b. bacteria
   c. decaying plants

③ Soil that has been transported and deposited by moving water is called
   a. fertile soil
   b. alluvial soil
   c. eolian soil

④ Fine soil particles like silt and clay that have been deposited by the wind create soil known as
   a. eolian soil
   b. alluvial soil
   c. loess soil

⑤ In many of the mountainous regions of the world, soil was formed by the action of
   a. volcanoes
   b. earthquakes
   c. neither a nor b

⑥ Fertilizers provide plants with
   a. insect repellent
   b. nutrients
   c. food

⑦ Plants that host rhizobia are called
   a. legumes
   b. fungi
   c. nematodes

⑧ One-celled organisms that live in moist soil are called
   a. protozoa
   b. nematodes
   c. fungi

⑨ Plantlike organisms that contain no chlorophyll are called
   a. protozoa
   b. legumes
   c. fungi

⑩ The area of the soil that contains the roots of plants is called the
   a. subsoil
   b. rhizosphere
   c. C horizon

Discussion

1 Soil is composed of what materials?

2 Why is plant life dependent on soil?

3 How is soil classified?

4 List two ways soil can be deposited.

5 What is loam soil?

6 What is the problem with too much clay in soil?

7 What are soil horizons?

8 What is the carbon cycle?

9 What is a symbiotic relationship?

10 What are the three basic groups of nematodes?

Plant Structures and Their Uses

Student Objectives

When you have finished studying this chapter, you should be able to:

- Describe the functions of plant stems.
- Tell the functions of xylem and phloem.
- Explain how stems are used by people.
- Describe the functions of a leaf.
- Explain how leaves are used by humans.
- Discuss the function of flowers.
- Explain how flowers are used by people.
- Explain how bees are used in agriculture.
- Describe the functions of seeds.
- Explain how seeds are used by people.
- Discuss the functions of roots.
- Tell how roots are used by people.

Key Terms

stems	flowers	seeds
photosynthesis	nectar	roots

All life as we know it on Earth is dependent on plants. Without plants, the atmosphere of our planet would not have the oxygen required to support animal life. As well as providing air to breath, plants are essential for food, fiber, erosion control, and many other necessities. Almost all of agriculture is dependent on plants. Plants provide food for humans and the agricultural animals raised for food (Figure 4–1). Plants, like animals, have different parts that serve different functions. People make use of these plant parts in a variety of ways.

Plants are made up of stems, leaves, flowers, seeds, and roots. Each of these parts plays an important role in the life of the plant. Agricultural producers grow particular plants in order to obtain plant parts or structures that can be used.

Stems

Stems are very important to plants. They support the plant and provide a means of moving nutrients from one end of the plant to the other. Some stems grow into large trunks, as in trees. Other stems may grow in the form of twining vines that allow the plant to climb fences or other plants.

FIGURE 4–1
Plants provide food for humans and feed for animals raised for food.
Courtesy of James Strawser, The University of Georgia.

At the very tip of the stem is a small knob called the terminal tip containing the bud that will become next season's growth. Further down the stem there are small scars that go all the way around the stem. These mark the place on the stem where last year's terminal tip or bud was. These scars are called bud scale scars. It is easy to calculate how old a stem is by counting these scars (Figure 4–2).

Every place along the stem where leaves are or were has scars, too. These are called leaf scars and they occur at nodes. Also at the nodes are small buds called axillary buds. These buds may at some time grow into stems or side branches. They grow in the axil formed by the leaf and stem (Figure 4–3).

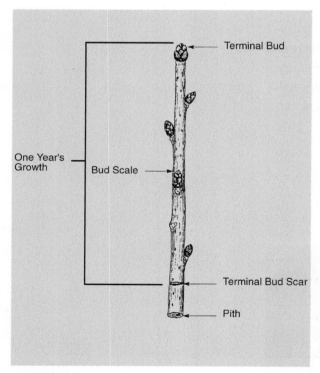

FIGURE 4–2
The age of a stem may be calculated by counting the bud scale scars.

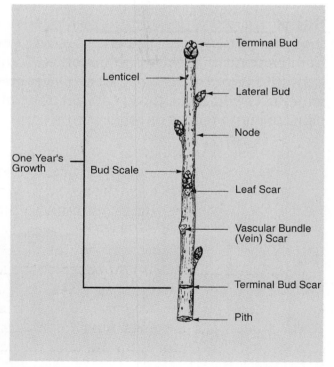

FIGURE 4–3
Parts of a stem.

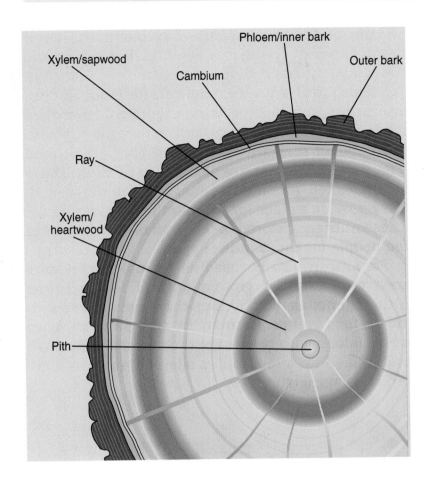

Phloem/inner bark

Xylem/sapwood

Outer bark

Cambium

Ray

Xylem/
heartwood

Pith

FIGURE 4–4
The xylem of trees is used to make paper.

Inside the stem are tubelike structures called the xylem and the phloem. The xylem carries the water and nutrients absorbed by the roots to the leaves. The phloem carries sugars manufactured in the leaves down to the roots. Agriculture makes use of these parts. In trees, the woody part is the xylem. Lumber cut from trees is composed of the xylem of the stems. Power poles, fence poles, and plywood are composed of stem materials. Also, the xylem of stems from trees provides fiber for making paper (Figure 4–4). Just think of all the uses for paper and paper products.

In a plant called flax, fibers are removed and woven into a cloth called linen. These fibers are the

FIGURE 4–5
Stems of plants like asparagus are used for food. *Courtesy of James Strawser, The University of Georgia.*

phloem of the stems of the flax plant. The outer layers of woody tissue (the xylem) are removed and the phloem is harvested.

Stems are also used extensively for food. The celery we eat in salads or as an ingredient in cooked foods is a stem. Rhubarb and asparagus are also examples of stems that are used for human food (Figure 4–5). Potatoes are also a type of stem. These stems, called modified stems, grow underground. They provide a place for the plant to store nutrients. These nutrients, stored in the form of starches, have fed people for thousands of years.

Stems can also be used as animal feed. Plants such as alfalfa are ground up and fed to cattle. Although not as nutritious as the leaves, stems provide a food source. Corn stalks and grain sorghum stalks (along with the rest of the plant) are chopped and made into a feed called silage. Many cattle are fed on silage (Figure 4–6).

FIGURE 4–6 Cattle eat plant stems in the form of silage. *Courtesy of James Strawser, The University of Georgia.*

Leaves

Everyone is familiar with leaves. They provide the beautiful green colors of spring and the reds, yellows, and browns of autumn. Leaves add color and decoration both inside and outside our homes. Potted plants with unusual leaf colors are used inside the home for decoration. Plants that turn different shades and different colors with the seasons are used outside.

Leaves also come in a wide assortment of shapes and sizes. Size, shape, and patterns of leaves are used to identify plants. The leaves of a juniper are less than ¼ inch long, but the leaves of elephant ear plants can be as large as 3 feet across (Figure 4–7). Leaves grow in alternate, opposite, or whorled patterns. In the alternate arrangement, leaves appear on one side of the stem at a node and from the other side at the next node. Opposite leaves grow from opposite sides of the stem directly across from each other. Whorled leaves refer to three or more leaves that grow around the stem at the same node (Figure 4–8).

Leaves are where the manufacture of food takes place. In a process known as **photosynthesis**, the energy of sunlight is used to produce nutrients. These nutrients are used by the plant for growth,

FIGURE 4–7
Some plants like the elephant ear have huge leaves.

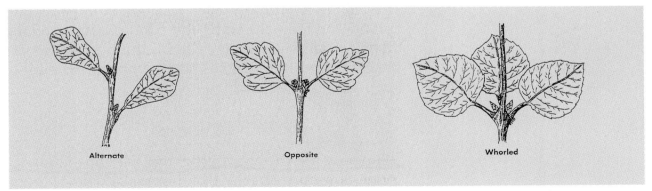

Alternate   Opposite   Whorled

FIGURE 4–8 Leaves of different plants are arranged differently on the stem.

FIGURE 4–9
Humans also make use of plant leaves for food. *Courtesy of James Strawser, The University of Georgia.*

maintenance, and reproduction. This food is in the form of sugars and carbohydrates as well as fats and proteins. Because this food is stored in leaves, animals use the leaves for food. Over the years, agriculturists have discovered and developed plants with leaves that are high in nutritive value and that livestock will eat. Cattle, sheep, and horses eat the leaves of grass and other plants as forage. Producers cut and dry these forages into hay that is stored and used as livestock feed.

Humans also eat the leaves of plants (Figure 4–9). As with plants used to feed animals, scientists have developed those plants that make good food for humans. Cabbage, mustard, lettuce, and spinach are all examples of plant leaves eaten by humans. The leaves are a primary place for food storage in the plant. The food stored by the plant can be used as food for animals. Many of the nutrients needed by animals and humans are obtained from the eating of plant leaves.

Flowers

Flowers are the more attractive part of the plant. Most flowers are flashy, like the rose or daisy. Others are so plain and simple, it is hard to realize they are flowers. Flowers serve as the reproductive organs for many plants. Some plants reproduce asexually. This means that these plants reproduce without a sperm and egg uniting. Usually, these type plants do not have flowers. Plants that reproduce sexually have flowers.

Some species of plants have male flowers and female flowers on one plant. Others have female flowers on one plant and male flowers on another plant. Most, however, have both the male and the female parts in the same flower.

Most flowers consist of the same basic parts. The female part, called the pistil, is made of the stigma, style, and ovary. The male part is called the stamen and contains the anther and filament. The pistil produces eggs that are fertilized by the pollen produced in the stamen. Pollen is usually taken from the anther to the stigma by insects, the wind, or in some cases, animals (Figure 4–10).

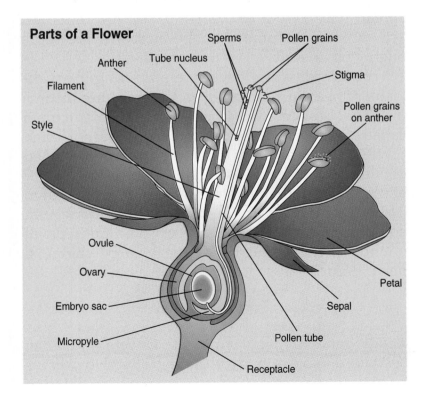

Parts of a Flower

Sperms

Pollen grains

Tube nucleus

Anther

Stigma

Filament

Pollen grains on anther

Style

Ovule

Ovary

Petal

Embryo sac

Sepal

Micropyle

Pollen tube

Receptacle

FIGURE 4–10
The parts of a flower.

FIGURE 4–11
Brightly colored flowers attract insects that pollinate the plants. *Courtesy of James Strawser, The University of Georgia.*

FIGURE 4–12
Honeybees are particularly adept at pollinating. *Courtesy of James Strawser, The University of Georgia.*

In order to attract the bees or other insects, plants have developed brilliantly colored flower petals (Figure 4–11). Although insects probably don't see the colors of the flowers the same way as we do, they are attracted by the petals. The flowers secrete a sweet fluid known as **nectar** that the insects use for food. As they crawl or fly into a flower to collect nectar, grains of pollen are caught on their legs or other body parts. When they go to the next flower, the grains of pollen are knocked off and remain in the flower. This causes pollination to occur.

Honeybees are particularly adept at pollinating (Figure 4–12). Many insects work flowers, but most go from one kind of flower to different kinds of flowers each day. Bees, on the other hand, work a particular kind of flower for a period of time. For example, honeybees will work apple blossoms for several days until the flowers are gone then work a different type of flower. By doing this they go from an apple blossom to another apple blossom and spread pollen from one apple blossom to another. This process ensures that the blossoms are thoroughly pollinated.

Fruit growers hire beekeepers to bring in truckloads of bees in the spring as the trees are blooming. The bees live in wooden, boxlike structures called hives with a separate colony of bees in each hive. The hives are easy to handle and can be loaded on a truck with the bees still in the hive. The owner of the bees can then move the hives from orchard to orchard for a fee from the fruit or crop producer. In addition, the producer can harvest hundreds of pounds of honey each year that can be sold at a profit (Figure 4–13).

Seeds

Seeds generally result from the flowers on plants. Seeds are another part of the reproductive process of the plant. Most plants begin their growth from seeds. The production of seed is one of the most important links in the entire agricultural industry. In fact, a lot of our food is derived from seed. Estimates are that around 70 percent of our diet comes directly from the seeds of plants. The bread or bowl of cereal that you had for breakfast this morning was made from the seed of grain plants (Figure 4–14). Cooking oil is derived from the oil that is pressed from such seeds as safflower, soybean, or cotton seed. Snack foods such as peanuts and popcorn are all made from the seeds of plants.

Seeds are also used for a variety of other purposes. Many of our beverages, such as coffee and cocoa, come from the seeds of trees and shrubs. Oils and dryers in paints and wood finishes come from seeds. Oil from the seed of the castor bean was once used for medicine and is now used as an oil in hydraulic systems. Seeds are also used in the manufacture of cosmetics, ointments, and pharmaceuticals.

Seeds produced on plants often are encased in a fruit. For example, apple seeds develop in the pulpy sweet material that is eaten by people and animals

FIGURE 4–13
Commercial honey producers place bees in an orchard to pollinate the trees and to produce honey. *Courtesy of Keith Delaplane, The University of Georgia.*

FIGURE 4–14
Breakfast cereals are made from plant seeds. *Courtesy of James Strawser, The University of Georgia.*

FIGURE 4–15
The fruit of an apple encases seeds. *Courtesy of James Strawser, The University of Georgia.*

(Figure 4–15). Many different fruits encase seed. Plums, peaches, pears, and oranges all have seeds at the center. Some foods we don't generally think of as fruits also fit this description. Tomatoes, squash, eggplant, and watermelons all encase seed and are eaten by people.

Roots

The two main functions of **roots** are to support the plant and to take in water and nutrients from the soil. Plant roots are at least as extensive as its stems and leaves. When you see a large tree just think of how much of it is really in the ground, hidden from your view.

There are four main types of roots: fibrous, tap, adventitious, and aerial. Fibrous root systems have many tiny roots that lead from the plant deep into

Career Development Events

The Food for America program is an excellent way to expose your FFA to elementary school children. This program is designed to make young students aware of the products and opportunities of agriculture. Many elementary students have never had the opportunity to visit any agriculture industries, businesses, or farms. The agricultural source of many everyday products, from milk to shoe polish, is easily overlooked.

Few Americans have an accurate understanding of modern agriculture. Many of today's consumers believe that since the number of farmers is decreasing that agriculture is dying. What

FFA Members help elementary students understand where food comes from. *Courtesy of Blane Marable, Morgan Co. High School, Madison, GA.*

the ground. One corn plant can have as much as 100 miles of roots.

Most trees have taproots. These are very large tough roots that reach deep into the earth for moisture and keep the tree from falling over. Carrots are also examples of a taproot system. There are many smaller roots that come off the main root.

Adventitious roots usually occur on plants that also have fibrous or taproot systems. An example is corn plants that have roots coming out of the main stalk above the ground that help to brace the plant against strong winds.

Aerial roots are common in very wet areas of the world. These roots are able to obtain water directly from the air. Plants with this type of root system don't even need soil to grow in. Both orchids and staghorn ferns have aerial roots.

they fail to realize is that agriculture technology has helped farmers become more productive. Today's agricultural producer uses less land to provide for more people. The Food for America program exposes the network of the many industries associated with agriculture from production, transportation, and storage to the marketing of food and other agricultural products. Careers as scientists, marketing specialists, engineers, veterinarians, nutritionists, and other occupations are important parts of American agribusiness and should be discussed with elementary students. Through the Food for America program, FFA members help elementary students understand these crucial roles and how food and other agricultural products are produced. FFA members should help elementary school teachers in agricultural instruction and bring examples of animals and agriculture products into the classroom. FFA members may even offer tours of a working farm as a field trip for the elementary students. FFA members are excellent role models. With well-practiced presentations, FFA members can leave a positive image of today's agriculture and the FFA through the Food for America program.

FIGURE 4–16
Humans also make use of roots, such as sweet potatoes, for food. *Courtesy of James Strawser, The University of Georgia.*

As well as being a necessary part of the plant, roots are important to humans. Plants are often used to slow down or stop erosion. Roots help to hold soil in place. All of us have eaten roots at one time or other. Carrots, turnips, radishes, and sweet potatoes are all examples of roots that are commonly eaten (Figure 4–16). Some spices also come from roots. Ginger is a root that is grown in India and Southeast Asia. It has been used for hundreds of years to add spice to foods.

Summary

Plants have always been used by people to eat, build shelter, and make clothing. Almost all of agriculture is involved with plants. Without plants and their various uses, people would have a difficult time existing. Agriculture has improved plants that were once found in the wild. Chosen for a particular use, these plants were bred to be more efficient. Modern agriculture will continue to produce these plants to fill the needs of people.

CHAPTER REVIEW

Student Learning Activities

❶ Make a list of all the food in your house that comes from plants. Determine the part of the plant from which the food was made. Share the results with the class.

❷ Look around your home and list all the plants that are ornamental. What parts of the plants provide the beauty?

❸ Choose a particular plant part (such as leaves, stems, roots, or flowers) and create a list of all the different uses that could be made of this part.

④ Locate several different types of flowers. Dissect them and identify the parts.

⑤ Bring in at least five products made from seed. Report to the class on how seeds were used in making the products.

True/False

① ___ Potatoes are a type of stem.

② ___ Plants reproduce asexually.

③ ___ Leaves serve as the reproductive organs for many plants.

④ ___ Some species of plants have male flowers and female flowers on one plant.

⑤ ___ Nectar is a pesticide.

⑥ ___ Aerial roots are common in cold areas.

⑦ ___ Plants are often used to slow down or stop erosion.

⑧ ___ Whorled leaves refer to two or more leaves that grow around the stem at the same node.

⑨ ___ Leaves are where the manufacture of food takes place.

⑩ ___ In a process known as photosynthesis, the energy of sunlight is used to produce nutrients.

Multiple Choice

① At the very tip of the stem is a small knob called the
    a. terminal tip
    b. node
    c. xylem

② The structure inside the stem that carries the water and nutrients from the roots to the leaves is the
    a. phloem
    b. node
    c. xylem

3 Flowers are really just specialized
   a. leaves
   b. plants
   c. vegetation

4 Those plants that reproduce sexually have
   a. leaves
   b. flowers
   c. stems

5 Honeybees are particularly adept at
   a. asexual reproduction
   b. producing nectar
   c. pollinating

6 Most tree roots are
   a. fibrous
   b. tap
   c. aerial

7 The xylem of tree stems provides fiber for making
   a. cotton
   b. linen
   c. paper

8 The male part of the plant is called the
   a. pistil
   b. stamen
   c. stigma

9 Insects are attracted to the
   a. petals
   b. leaves
   c. color

10 Seeds produced on plants often are encased in
   a. plants
   b. soil
   c. fruit

Discussion

1. Why are plants so essential to us?

2. What are the major parts of plants?

3. What are the functions of the xylem and phloem?

4. What are some uses humans make of plant stems?

5. What are the parts of the flower?

6. Why are insects important to flowers?

7. List the main types of roots.

8. Why is the production of seed one of the most important links in the agriculture industry?

9. For what purposes can leaves be used?

10. How are bees used in agriculture?

CHAPTER 5

Agricultural Pests

Student Objectives

When you have finished studying this chapter, you should be able to:

- Explain why producers have to deal with pests.
- Discuss how insects are harmful.
- Define the life cycle of insects.
- Discuss how insects are controlled.
- Analyze the concept of integrated pest management.
- Explain what a weed is.
- Tell how weeds are controlled.
- Discuss the newest methods used in pest control.

Key Terms

insects	parasites	biological control
pests	insecticides	weeds
disease	integrated pest management	herbicide

Agriculture deals with the growing and cultivating of living things for the use of humans. People provide the best care possible for the plants and animals they grow. One of the basic parts of caring for crops and animals is protecting them from other living things. In the wild, plants and animals are constantly in competition with each other. In nature, all living things need space and nourishment. Animals eat other animals, animals eat plants, and plants get nutrients from the soil. When humans grow plants or animals, they provide an enticing meal for animals such as insects. Every year, cropland is prepared for planting. The soil is tilled, fertilizer is spread, and conditions are made favorable for the crop to grow. Unfortunately, the conditions also favor the growth of unwanted plants.

Many industries have developed to protect plants and animals from pests, such as the manufacture of cultivation equipment and the development of agricultural chemicals (Figure 5–1). These industries use a tremendous amount of resources each year. Machines and chemicals have to not only be made but also have to be distributed to the producer.

FIGURE 5–1
A vast industry has developed to protect plants from pests. *Courtesy of James Strawser, The University of Georgia.*

Insects

Insects are what most people think of when they think of agricultural **pests**. Since humans first began growing their own food, insects have been a problem. Many ancient documents tell of troubles with insects, such as locusts, that devoured crops. For thousands of years it was accepted that insects would get a large percentage of the crops (Figure 5–2). Problems with insects continue even to this day. If insects are not controlled on crops, millions of dollars' worth of damage will be done. Not only will insects eat a large part of the produce, but they will also damage most of the crop not eaten. We are accustomed to seeing crisp, fresh apples at the

FIGURE 5–2
In the past it was accepted that insects would get a large percentage of the crop. *Courtesy of James Strawser, The University of Georgia.*

grocery store. These apples are almost flawless and have no damage. However, if insects are not controlled as the apples are growing, the apples we buy will have defects such as worm holes. Most consumers would not buy this type of apple (Figure 5–3). The same is true of almost all that we grow.

Not only do insects eat our crops, but they also carry **disease**. Millions of people have contracted diseases because of insects. Insects such as the mosquito may draw blood from a person infected with a disease and then inject disease organisms in the next person it bites. If the same mosquito bites many people, the disease becomes widespread (Figure 5–4). Yellow fever, malaria, West Nile Virus, and encephalitis are diseases spread by mosquitoes. Throughout history, countless millions of head of livestock have died as a result of diseases carried by insects.

Some insects are **parasites**. Parasites are organisms that live on other organisms (called hosts) and harm them. Agricultural animals are harmed by insect parasites such as ticks, fleas, and heel flies. These insects get their nutrients from the animal they live on or in. This causes the animal to have to eat more in order to live and do well. In turn, the producer must give more feed to the animal or control the parasite.

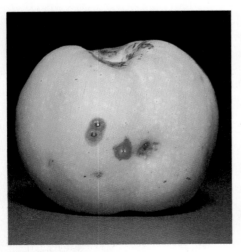

FIGURE 5–3
Consumers are reluctant to buy damaged fruit. *Courtesy of James Strawser, The University of Georgia.*

One of the biggest problems with insects is that there are so many of them. Over 750,000 species have been classified. Scientists think there may be as many as 2,000,000 more that have not been identified. Of the identified insects, only about 500 are considered to be major pests in this country. However, these pests are formidable enemies. One female insect may lay millions of eggs during her life (Figure 5–5). If most of these hatch out, there will be a lot of insects around to cause trouble. Fortunately, nature helps out. Many birds and animals eat large amounts of insects. Other insects such as the lady beetle and the praying mantis devour huge numbers of insects. Each year, vast numbers of insects are eaten by predators.

Even with all the predators eating insects, that still leaves a lot to destroy crops. For many years we have used chemicals to get rid of insects. These chemicals, called **insecticides**, have been very efficient. They have been the most efficient means ever devised to kill insects. In the past, insecticides have been used in tremendous quantities. The overuse of insecticides has resulted in problems. Many of the older chemicals such as DDT have been harmful to

FIGURE 5–4
Insects such as mosquitoes carry disease. *Courtesy of USDA ARS K 4705-01.*

FIGURE 5–5
One female insect may lay millions of eggs in her lifetime. *Courtesy of James Strawser, The University of Georgia.*

FIGURE 5–6
Reproductive cycles of animals such as water birds and eagles were disrupted by DDT. *Courtesy of Getty Images.*

the environment, and wildlife has suffered. These chemicals remain in the soil for many years. Rain would wash the chemicals into streams where they were absorbed by aquatic plants and animals. Animals such as water birds and eagles ate the fish and absorbed the chemicals. Reproductive cycles were disrupted because the eggs from the birds would not hatch (Figure 5–6).

Today, all insecticides are regulated very closely by governmental agencies. Most of the old insecticides that remained in the soil for so long have been banned. New insecticides break down into harmless substances in a short time. This makes environmental problems less likely because the residues of the chemicals do not build up.

Any sound insect control program involves **integrated pest management** (IPM). This approach combines knowledge of the pest and its life cycle, monitoring, cultural practices, biological control, and chemical pesticides.

Much has been done recently in the area of **biological control**. Biological control involves the use of predatory insects, fungi, or viruses. You are probably familiar with the appetite of insects such as the praying mantis and the lady beetle (Figure 5–7). In some areas, these insects are raised for the purpose of eating insects on crops.

Cultural practices have a big impact on diseases and insects. Simply pruning out dead and diseased limbs makes a difference. Burning these diseased limbs and weeding around plants helps a lot. This helps to break up the life cycle of insects. Insects go through several phases in their lives (Figure 5–8). The cycle begins when the female lays eggs. The eggs hatch into wormlike organisms called larvae. Most of the harm insects do comes about during this stage because the larvae eat a tremendous amount of plant material. Larvae develop into the pupa stage and may be covered for a period by a

FIGURE 5–7
Insects, such as the lady beetle, devour other insects. *Courtesy of USDA ARS K 4249-04.*

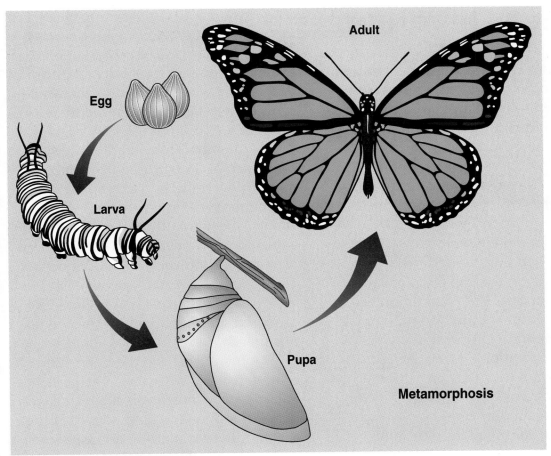

Egg

Larva

Adult

Pupa

Metamorphosis

FIGURE 5–8 Insects go through a life cycle.

cocoon. The pupa then develops into the adult stage. The adult lays eggs and the cycle begins all over again. Depending on the type of insect, they are more vulnerable at one phase than others. For example, if they spend the winter as a pupa in crop residue or in dead fruit tree limbs, getting rid of the crop residue or burning the dead limbs will get rid of the insects.

In integrated pest management, a combination of methods is used to control the insects. The crop is closely monitored and chemicals are used only when necessary. This not only cuts down on production costs but also helps protect the environment. ⌉ Stop!

Career Development Events

Agriculture students should be aware of the damage that insects, diseases, and weeds create in all areas of agriculture. The combination of classroom study and FFA activities prepares students to handle pest problems correctly and safely. Agriculture teachers educate students on methods of identification, control, and prevention of these pests. The FFA offers students the opportunity to compete and be awarded for their knowledge on this topic.

The National FFA Foundation provides funds for recognition of members' achievements at the local, state, regional, and national levels. Awards to members are generally presented at the annual chapter banquet. Competition includes team events and individual efforts to identify plant pests and disorders in forestry, floriculture, and nursery/landscape career development events. When students judge plants, they need to be able to look for symptoms of insect, parasite, and disease damage and presence.

FFA members may use a pest-related topic such as "Integrated Pest Management in Agriculture" to compete for the FFA public speaking career development events. Speaking out on agriculture issues not only develops skills in communication but informs the public about changing times. There are two FFA public speaking contests—a 6- to 8-minute prepared speech and an extemporaneous speech.

FFA members have the opportunity to research and write prepared speeches. The speech manuscript is submitted to a panel of judges and used as part of the judging. The extemporaneous speech competition is designed to test the FFA member's ability to talk about a given

Weeds

Weeds are plants that are growing where they are not wanted. If oats are growing in a wheat field, the oats may be considered to be weeds. In fact, some of our worst weed pests were once considered to be good plants. One of the best examples is Johnsongrass, which was brought in from Africa. In the 1800s, it was thought that this grass would be the perfect pasture grass in the South. It is a

subject without having prepared or rehearsed its content. The FFA member is given 30 minutes to generate a 4- to 6-minute speech on a given agriculture topic. Both events judge the student's power of expression, stage presence, voice control, knowledge, and response to questions.

In several FFA Career Development Events, students must inspect plants for insect damage. *Courtesy of Carol Duval, National FFA Organization.*

fast-growing grass that can be used for grazing. However, the grass got into other fields and competed with crops. It has now spread all across the country and is extremely difficult to kill in crops. It outgrows most agricultural plants and robs them of sunlight and nutrients (Figure 5–9).

Weeds are pests in many different ways. They can harbor insects and diseases. Remember that insects go through several life phases. They may need weeds as a means of getting through one of

FIGURE 5–9
In the 1800s, Johnsongrass was thought to be a good pasture grass. Now it is a serious pest. *Courtesy of William K. Vencill, The University of Georgia.*

the phases. Weeds compete with agricultural plants for nutrients and water. Weeds must have nutrients the same as agricultural plants. If a field is full of weeds, the weeds may get as much or more of the soil's nutrients as the good plants. This means there are fewer nutrients for the crop. Also, the weeds may block out the sunlight (Figure 5–10). Sunlight is needed by the plants to produce sugars in the process of photosynthesis.

Another problem with weeds is that they may lower the quality of the crop. Hay that has weeds in

FIGURE 5–10
Weeds can block out sunlight from reaching other plants. *Courtesy of William K. Vencill, The University of Georgia.*

it may be less nutritious or may even contain poisonous weeds. Grain that contains weed seed is not as good for human food or animal feed. A cotton crop that has weeds will have foreign particles in the lint and will be less valuable.

Most weeds are hardy, produce many seeds, and can completely take over an area very quickly (Figure 5–11). The fact that weeds are still around indicates that they are hard to control. People have been fighting them for thousands of years and the fight goes on.

People have used many means to fight weeds. Pulling the weeds up, hoeing, and plowing are the oldest means. Also, animals have been used to get rid of unwanted plants. The animals are brought in and confined to an area where they eat the weeds. The most efficient way of controlling weeds is the use of a **herbicide**, a substance that kills plants. Chemical herbicides have been widely used since the 1960s and continue to be a very effective means of weed control.

There are basically two types of herbicides. One is the nonselective herbicides. This type kills all plants that it contacts. These are used in places where all of the vegetation is considered to be weeds. For example, nonselective herbicides are used on railroad beds to kill all plants. They also can be used with crops if the herbicide only contacts the weeds and not the crops.

The other type is selective herbicides. These will kill only certain types of plants (Figure 5–12). For example, a selective herbicide might kill cockleburs that are growing in corn. The herbicide will not affect the corn plants even if it comes in contact with them.

Years of research have gone into developing selective herbicides. A new herbicide must pass several requirements. First, it must not harm the crop it is used on. Second, the chemical has to be safe to

FIGURE 5–11
Weed plants usually produce a lot of seed. *Courtesy of James Strawser, The University of Georgia.*

FIGURE 5–12
Selective herbicides kill only the weeds. *Courtesy of James Strawser, The University of Georgia.*

FIGURE 5–13
Pesticides have to be safe to apply.
Courtesy of James Strawser, The University of Georgia.

use by producers. Producers come in contact with the herbicides as they are applied. If the herbicide is toxic to humans, the applicator might be harmed. Third, it must not be harmful to the environment. We have to protect plants and animals around us, so we have to be careful that the herbicide kills only the plants we intend. Fourth, the chemical must be affordable to use. It would do no good to use a chemical that was more expensive to use than the benefit produced. Modern herbicides must fit all of these requirements before they are released for use (Figure 5–13).

Some of the newer means of controlling weeds include the use of biological agents. This means that organisms such as insects and disease germs can be used. If an insect's favorite food is a particular weed, the insect might be used to control the weed. Of course, researchers have to be careful that the insect does not develop an appetite for crops, too. If the insect meets the requirement of only eating the weed pest, it can be raised for the purpose of eating the weed.

Diseases also can be used to fight weeds. Some diseases are particular about the type of plant they

FIGURE 5–14 Diseases are a serious problem in plants. The plant shown in (A) is healthy; the one in (B) is diseased. *Courtesy of James Strawser, The University of Georgia.*

attack. Generally, a particular bacteria, virus, or fungus will attack only a certain type of plant (Figure 5–14). If a disease-causing agent can be found that only attacks a particular weed, the disease might be effective in controlling the weed.

Genetic Engineering

One of the most exciting possibilities for controlling agricultural pests is the use of genetic engineering. Genes are the substances in the cells of plants and animals that contain the material (DNA) responsible for passing characteristics from one generation to the next. Scientists now have the ability to take genes from one organism and place them in another. This means that they can alter the characteristics of plants and animals. Plants have been developed that are resistant to insects. This means that insects will not be as likely to feed on the genetically altered plants. Crops known as Bt crops are now common. These crops, such as corn and cotton, have been genetically engineered to produce a toxin from a soil bacteria, *Bacillus thuringiensis* (Bt). This toxin kills insects but is not harmful to humans or other mammals.

We now have the ability to make plants more tolerant to herbicides. Obviously, the more tolerant a crop plant is to a herbicide, the more likely weeds can be controlled. One of the problems is that the more similar the crop is to the weed, the more difficult it is to use a selective herbicide. By using genetic engineering, we can make the crop more unlike the weed.

Summary

Humans have always faced pests that threaten agriculture. Although we spend a lot of time and money growing crops and animals, they are still a part of nature. In nature, plants and animals depend on each other. It is a difficult task to keep agricultural plants and animals away from the other organisms in nature. Through study and experimentation, we will better understand agricultural pests and find new and better ways of controlling them.

 CHAPTER REVIEW

Student Learning Activities

❶ Collect 10 insects that are harmful and 10 that are not harmful. Identify the insects and tell what damage the harmful insects do.

❷ Identify a harmful insect that is common to your area. Devise a means of controlling the insect. Make sure your plan uses the integrated pest management concept.

③ Identify the weeds that are common to your area. Collect five and bring samples to class. Discuss with the class how the weeds spread.

④ Locate plants that have a disease. Try to identify the disease.

True/False

① ___ Insects do not carry diseases.

② ___ Agriculture deals with the growing and cultivating of living things for human use.

③ ___ One of the basic tasks in caring for crops and animals is protecting them from other living things.

④ ___ Cancer and diabetes are diseases spread by mosquitoes.

⑤ ___ Some insects are parasites.

⑥ ___ Diseases also can be used to fight weeds.

⑦ ___ One of the biggest problems about insects is that most are deadly to humans.

⑧ ___ Insecticides are the most efficient means ever devised to kill insects.

⑨ ___ Any sound insect control program involves integrated pest management.

⑩ ___ Weeds are plants that are growing where they are not wanted.

⑪ ___ Genetic engineering is used in insect control.

⑫ ___ Almost all insects are harmful.

Multiple Choice

① The insect life cycle begins when the female
   a. dies
   b. mates
   c. lays eggs

2 Insect eggs hatch into wormlike organisms called
   a. larvae
   b. pupas
   c. cocoons

3 A substance that kills plants is called a
   a. pesticide
   b. herbicide
   c. insecticide

4 Organisms that live on other organisms and harm them are called
   a. insects
   b. parasites
   c. predators

5 Many people have contracted diseases because of
   a. insects
   b. plants
   c. predators

6 An estimation of the number of major insect pest species in the United States is
   a. 750,000
   b. 2,000,000
   c. 500

7 Two insects that devour huge numbers of insects are
   a. lady beetles and praying mantises
   b. ticks and fleas
   c. cockleburs and honeybees

8 Johnsongrass is considered to be a
   a. landscaping grass
   b. perfect pasture grass
   c. weed

9 Plants have been developed that are resistant to
   a. insects
   b. pesticides
   c. animals

⑩ Every year cropland is prepared for
    a. animals
    b. insects
    c. planting

Discussion

① List and discuss some of the problems caused by insects.

② Define integrated pest management.

③ What is meant by the biological control of insects?

④ Why are weeds considered pests?

⑤ What is the difference between a nonselective and a selective herbicide?

⑥ How can genetic engineering be used for controlling agricultural pests?

⑦ How do cultural practices have an impact on diseases and insects?

⑧ What is the life cycle of an insect?

⑨ How can weeds be controlled?

⑩ List two types of biological agents.

CHAPTER 6

Floriculture

Student Objectives

When you have finished studying this chapter, you should be able to:

- Define floriculture.
- Describe the national aspects of floriculture.
- Describe the international aspects of floriculture.
- Describe the three basic types of living plants florists sell.
- Describe the steps in producing a typical flower crop.
- List and describe the four flower arranging principles.

Key Terms

floriculture

bulbs

florists

potted plant

foliage plants

house plants

cut flowers

photoperiodic

FIGURE 6–1
Floriculture is a big agricultural business. *Courtesy of Washington State University.*

The word **floriculture** literally means the growing of flowers. Floriculture is a very big business in the United States and the world (Figure 6–1). Not only are billions of flowers produced each day, but also they must be taken to their destination, arranged, stored, and sold. Each part of this chain is encompassed by the floriculture business.

[handwritten: operations involved in]

International Aspect

Flower crops have been grown for centuries in Europe. Throughout history, flower gardens have produced beauty. Almost everyone enjoyed growing or looking at the flowers. Today, huge flower markets sell flowers by lots of hundreds and thousands. People seem to enjoy flowers as much as ever.

Most of the flowers brought into the United States come from Central and South America. The climate is favorable to the production of flower crops, and countries in the southern hemisphere can grow flowers during our winter. These are then shipped to the United States. Also, the cost of labor is low compared to other areas.

Other countries also send flowers to the United States. Australia and New Zealand sell their orchids here. Holland has always been famous for brilliantly

FIGURE 6–2
Tulips grown in Holland are shipped all over the world. *Courtesy of Jeff Lewis, The University of Georgia.*

colored tulips. Each year Holland ships out not only tulip flowers but also tulip **bulbs** (Figure 6–2). Israel has begun producing roses and is rapidly becoming a leading source for these flowers.

In return, the United States ships gladiolus all over the world. The fresh flower industry is a world-wide business. Many countries have close ties to each other in the floriculture business.

National Aspect

As with the international aspect, flower production in the United States is concentrated in areas where the climate is best suited. Most floral crops need very bright light to grow and produce quickly. Also, some flowers need a cooler climate and others need a warm climate.

The most popular states for growing flowers are Florida and California because of the warm climates. Colorado and the New England states are also heavily involved in producing flowers that require cooler climates. Bulbs, such as tulips and daffodils, need slightly different production climates. Cool summers and mild winters make Washington, Oregon, and California more suitable for producing these (Figure 6–3).

FIGURE 6–3
Some flowers need a cool climate, and others need a warm climate. *Courtesy of Getty Images.*

Plants for Florists

Many different kinds of plants are produced for **florists**. Everything from blooming plants in pots and hanging baskets to cut flowers such as roses and daisies are available to florists. This is a growing industry in this country. People buy flowers for weddings, funerals, anniversaries, birthdays, and other occasions. People also send flowers to cheer up friends who are sick or just to let them know they are special (Figure 6–4).

Potted Plants

Any plant grown and sold in a pot is referred to as a **potted plant**. Outside plants such as azaleas and chrysanthemums are grown in pots for sale at the florists. These are usually removed from the pot and transplanted outside. Blooming house plants like kalanchoe and cactus are sold in pots and generally remain in the pots.

Potted plants offer a challenge to the florist. Not only must they have beautiful or unusual foliage or

FIGURE 6–4
Creating floral arrangements is a growing business. *Courtesy of USDA.*

FIGURE 6–5
Blooming house plants are sold and live in pots. *Courtesy of James Strawser, The University of Georgia.*

blooms, but the plants have to be healthy. A potted plant, in most cases, must live, bloom, and remain attractive for a long time (Figure 6–5).

Most potted plants are grown from cuttings, while others are started from seed. It can take several weeks to a few months to produce a healthy, blooming potted plant.

Foliage Plants

Plants grown for their leaves are called **foliage plants** (Figure 6–6). Many times these are grown in pots and could also be called potted plants. Some plants, however, are grown for foliage for flower arrangements.

The most common plant for this use is the leatherleaf fern. Its fronds (leaves) are particularly strong. In many arrangements, leatherleaf accounts for most of the plant material. It is used to hide the support structure of the arrangement and as a green backdrop for the colorful flowers.

Another kind of fern, the Boston fern, is popular for hanging baskets. Other plants often used for hanging baskets are begonias and impatiens.

FIGURE 6–6
Foliage plants are grown for their attractive leaves. *Courtesy of James Strawser, The University of Georgia.*

Often several small foliage plants are put together in a shallow pot and sold as dish gardens. Groupings of plants inside glass or plastic containers are called terrariums (Figure 6–7).

The most popular foliage plants, however, are those grown for homes and public buildings. These can be sold in small pots but are also offered in very large sizes. Landscapers use these to make the yards and outsides of homes, businesses, and other structures more attractive.

Exotic Plants

Unusual or rare plants are called exotic plants. Many exotic plants are great for **house plants**. These are attractive because they are so unusual.

An example of an exotic plant is the bromeliads. These plants are unusual because of their cup-shaped leaves. The leaves act as cups because they hold water. For this reason, they can stand longer periods of neglect. Another example is the carnivorous plants. These are plants that eat flesh to grow. The Venus flytrap is the most common carnivorous plant. It traps flies and other insects and digests them (Figure 6–8). Another carnivorous plant is the pitcher plant. It traps insects by luring them into its

FIGURE 6–7
Plants can be used in a variety of ways to create beauty. *Courtesy of James Strawser, The University of Georgia.*

FIGURE 6–8
The Venus flytrap is an exotic plant that catches insects. *Courtesy of James Strawser, The University of Georgia.*

"throat." The sides are so slippery the insect cannot get out. It then falls into the digestive fluids in the bottom of the "pitcher" and dies. There is a good market for these exotic plants. People buy them to decorate their homes.

Career Development Events

The FFA Floriculture Career Development Event is a competition for teams of four members. Each member competes individually and the final team score is the total of the three highest individual scores. The event consists of four parts: (1) identification of plant materials, (2) placing classes, (3) general knowledge, and (4) practicum.

The identification of plant materials consists of plants selected from the State Floriculture list and is displayed for visual identification. Students may not touch the live plant material but can use a magnifying glass to inspect the foliage. The judging classes are plant materials or floral arrangements. The student places the materials from best to worst based on structural form, market value, and health. Floriculture plants are selected from the identification list and may be cut flowers, foliage plants, bedding plants, or potted plants. The general knowledge part of the floriculture event consists of 50 multiple-choice questions. This phase of the event tests knowledge and understanding of all areas of the floriculture industry. Students preparing for this part of the event must study not only the operations and maintenance of a greenhouse but also the physiology,

The Floriculture Career Development Event tests students' ability to work with floral arrangements. *Courtesy of Carol Duval, National FFA Organization.*

Cut-Flower Production

The main items in the floriculture trade, of course, are flowers. **Cut flowers** are flowers that are to be used in flower arrangements. They are cut at the

growth habits, nutritional needs, disorders, and propagation practices of floriculture crops. Questions addressing floral design, arrangements, equipment, and principles are included as well as customer service, interpersonal relations, salesmanship, and records and reports.

The part of the event known as the practicum is where the student identifies plant disorders. Plant disorders include disease, insects, weeds, and nutritional problems. The disorders are displayed for the student to view with or without a magnifying glass. The student is asked to identify the problem and those pesticides or other treatments necessary to remedy the problem. Hands-on practicums are included in the event. Students are asked to perform different skills required in the floriculture industry such as potting plants, designing a floral arrangement, or taking a phone order.

After all team members have completed each phase of the event, each individual score is calculated. The top three individual scores are added to yield the team score. If there is a tie between two team scores, the written

exam score is used as a tiebreaker for first and second place teams. State winning teams advance to the National Floriculture Career Development Event.

The FFA Floriculture Proficiency Award is one of the four horticulture proficiency awards. The award is presented to individual FFA members who have excelled in experiences using the principles and practices of field or greenhouse production of flowers, foliage, and related plant materials for ornamental purposes. To apply, the student must be involved in a Supervised Agricultural Experience Program and have maintained a list of floriculture skills he or she has learned. The student must also develop an inventory of buildings, equipment, tools, and materials used in the experience program. He or she must also keep accurate records of income, expenses, and accomplishments. Students may compete within their chapter, district, and state. The state award winners compete for national recognition.

FIGURE 6–9
Cut flowers are trimmed so they can fit into a vase. *Courtesy of Frank Flanders, The University of Georgia.*

FIGURE 6–10
In a greenhouse, flower growers can control the environment. *Courtesy of James Strawser, The University of Georgia.*

stems so they can fit into vases and other containers (Figure 6–9). This differentiates cut flowers from flowering plants. Flowering plants may be sold as potted plants. The plants that produce cut flowers are not sold. Only the flowers and stems are marketed.

Many nurseries produce flowers just for florists. Many produce only one kind of flower, such as chrysanthemums or carnations. For the growers to make a profit from their crops, they must practice good growing habits and timing.

Physical Environment

Although some cut flowers, such as gladiolus, are often grown outside in a field, most are grown in greenhouses. There are several reasons for this.

First is temperature. To induce plants to form buds, then flowers, they must have a certain range in temperature. Cold temperatures will keep many plants from forming buds at all. High temperatures may cause them to abort buds and blooms. Temperature can be controlled inside a greenhouse.

Another important factor is moisture. Plants for cut-flower production must have just the right amount of water to do their best. Producers can use sprinklers or underground water to keep plants in the field watered. However, it is difficult to prevent the plants from getting too much rain. Hard rain can also destroy the fragile blooms. Just as rain storms can damage plants, so can wind storms. By growing the plants in the controlled environment of a greenhouse, these conditions can be avoided (Figure 6–10).

Greenhouse Benches

Benches in greenhouses used for growing potted plants are usually from 5 to 8 feet wide. Benches used for the growing of cut flowers are only 4 feet wide. This is helpful for the intense care these

plants need (Figure 6–11). Plants used for cut flowers are usually not grown in pots. Instead, the
benches hold soil and the plants are grown in this.
These benches are also closer to the ground. Often,
flower stalks must be staked or tied to produce
straight stems, especially with the large flowers.
Lower benches allow more room in which to work.

Growing Cut Flowers

The grower usually starts with a cutting. Let's follow
a typical crop of chrysanthemums. Some growers
make their own cuttings, root them, grow them out,
induce blooming, harvest the flowers, and ship
them. More often, they start with rooted cuttings
(Figure 6–12).

   This is where timing comes in. Growers usually
have several crops growing at once so they can have
continuous production. A new crop is planted every
two to four weeks. Different types of chrysanthemums mature at different times, so the grower
must take this into account also.

FIGURE 6–11
Greenhouse benches hold pots for
growing plants. *Courtesy of James
Strawser, The University of Georgia.*

FIGURE 6–12 Some growers start new plants by taking cuttings as shown in (A) and doing rootings
as shown in (B). *Courtesy of James Strawser, The University of Georgia.*

FIGURE 6–13
Pinching makes the plants produce more limbs. *Courtesy of James Strawser, The University of Georgia.*

The cuttings are planted in the low, narrow benches and grown out. Wire mesh, usually with 6-by 8-inch spacing, is laid on top of the soil before planting. This is raised as the plants grow to assure straight stems on the flowers.

About two weeks after planting the plants are pinched. This refers to the practice of taking off the bud at the very tip of the plant. The plant will then produce several stems from buds farther down the stem (Figure 6–13). All but two or three of these stems are taken off. This makes the plants produce more flowers.

Lighting

Many plants such as chrysanthemums are **photo-periodic**. This means that they respond to the length of night. Only during long nights, or short days, do chrysanthemums produce flower buds. Left alone, they flower only during the fall and early winter. To get around this, the dark period is artificially controlled. To allow the plants to grow for a while without forming buds, the dark period is interrupted with electric lights.

When the plants are larger and the stems are long enough, lighting is no longer as important. If the greenhouse is in a rural area, without outside light coming in and it is the right time of year, the plants will form flower buds naturally. Most of the time, however, this is not left to chance. The plants are covered each evening with a solid piece of black cloth that blocks all light (Figure 6–14). The cloth is then removed in the morning. It only takes a very small amount of light from almost any source to interrupt the cycle. This continues until the flowers are well developed.

Temperature

Temperature must be monitored frequently. Although different cut-flower plants need slightly

FIGURE 6–14
A cloth over the greenhouse can help control the amount of light the plants get. *Courtesy of James Strawser, The University of Georgia.*

different temperatures, chrysanthemums need a night temperature of 60 to 62 degrees and a day temperature of 65 to 70 degrees.

If temperatures go above 80 degrees or so, it takes longer for flowers to be produced and the flower quality is lowered. By contrast, if temperatures are too cool, chances of disease go up, production time is longer, and the flower size and stem length are both less.

Timing

The grower must follow a precise timetable. This timetable must be started from the date the flowers must be ready and worked backward to pinpoint when the plants should be started, when short or long days should be started, when the plants should be pinched, when fertilized, and when temperatures should be controlled.

Timing is more important with some flowers. For example, unlike chrysanthemums, which are used all year in flower arrangements, poinsettias are grown for and bought only at Christmas. Green poinsettias are unsalable in December. Poinsettias that are a beautiful red and green on December 26

FIGURE 6–15
Poinsettias require rigid schedules of lighting, watering, and temperature control. *Courtesy of James Strawser, The University of Georgia.*

are likewise useless. Rigid schedules of lighting, watering, and temperature are used to make them bloom at the right time (Figure 6–15).

Harvesting

Chrysanthemum flowers may be harvested when they are just opening or in full bloom. This depends on the type of flower and how far the flowers will travel to market. The stem is cut as long as it can be. This is to give the florist enough stem length for a variety of uses. Harvesting continues for one to two weeks. Then the plants are discarded and the bed is prepared for the next crop.

The flowers are immediately put into room-temperature water, then into a cooler where they are kept at a much lower temperature. Once the flowers are cut, they are no longer living and will begin to decay. Cool temperatures delay the decay until the flowers are used. Later, the flowers are graded. The best grade includes the largest, most perfect flowers (Figure 6–16). Smaller flowers or slightly off-color flowers are lower grades.

FIGURE 6–16
The best grade of flowers is large and perfect. *Courtesy of James Strawser, The University of Georgia.*

FIGURE 6–17
Flowers have to be carefully packed to prevent damage. *Courtesy of University of Minnesota Agricultural Experiment Station.*

Transportation

Flowers are easily damaged and special measures are taken to protect them (Figure 6–17). The bunches (for chrysanthemums there are usually 12 in a bunch) are wrapped in a clear plastic. Large, sturdy cardboard boxes are used to further protect the flowers. Newspaper is packed gently around the flowers to keep them from jostling.

Flowers are transported in refrigerated trucks for short distances. For longer distances, such as from California to Maine, or Colombia to Miami, they are shipped in refrigerated compartments in airplanes.

Some of the more delicate flowers only last a matter of days, then are unusable. For this reason, the florist must know when to order these.

Flower Arrangements

How well the flowers look depends to a large degree on how they are presented or arranged. Among the principles of flower arranging are balance, scale, focal point, and unity (Figure 6–18).

FIGURE 6–18
A nice floral arrangement must have a focal point, balance, scale, and unity. *Courtesy of James Strawser, The University of Georgia.*

Balance refers to the apparent weight of the flowers. An arrangement should not look like it is about to fall over. By balancing large flowers and light and dark flowers, the designer creates a beautiful piece that looks stable.

Scale involves the various sizes of flowers compared to each other and the container. It would not look right if the arrangement consisted of a few small flowers in a very large vase.

An arrangement should have something that draws the viewer's attention. This is called the focal point. The focal point is usually in the front, near the bottom of the arrangement. It can be a very large flower, a very dark flower, or an unusual flower.

The design should flow together. A wild flower bouquet with one rose would be out of place. Unity is necessary in an arrangement. All the parts should fit as one.

Summary

From growing flowers and plants to selling the finished product, floriculture is a worldwide business. It takes planning and forethought to produce and sell floriculture products.

CHAPTER REVIEW

Student Learning Activities

1. Bring a variety of different types of flowers to class. Arrange them in different ways. Try to identify several principles that apply to making floral arrangements that look good.

2. Interview a florist. Determine where the flowers used by the florist come from and the main markets for the arranged flowers. Share your findings with the class.

③ Obtain a potted plant from a department store, florist, or friend. Go to the library and conduct research about the plant. Find out the country where the plant originated and where the domesticated plants are grown. Also determine how the plants are produced. Report to the class.

④ Choose a type of flower such as roses, tulips, poinsettias, orchids, etc. Research the history of the flower. How far back in time have humans cultivated them? What countries developed them? What interesting facts can you discover?

True/False

① ___ Most potted plants are started from seed.

② ___ Floriculture is the growing of flowers.

③ ___ Plants grown for their leaves are called foliage plants.

④ ___ Photoperiodic means that the plants respond to the length of the day.

⑤ ___ If temperatures rise above approximately 80 degrees, it takes longer for flowers to be produced and the flower quality is lowered.

⑥ ___ Balance refers to the apparent weight of the flowers.

⑦ ___ The most common plant grown for foliage is the pitcher plant.

⑧ ___ Dish gardens are deep pots for small foliage plants.

⑨ ___ Unusual or rare plants are called exotic plants.

⑩ ___ Terrariums are plants growing inside a glass or plastic container.

Multiple Choice

① Most of the flowers brought into the United States come from
   a. China and Japan
   b. Central and South America
   c. Australia and New Zealand

② One of the leading sources for roses is
   a. China
   b. Israel
   c. Australia

3 Flower production in the United States is greatest in
   a. Florida and California
   b. Georgia and Florida
   c. North Carolina and California

4 The main items in the floriculture trade are
   a. shrubs
   b. plants
   c. flowers

5 Benches in greenhouses used for growing potted plants are usually between
   a. 5 and 8 feet wide
   b. 3 and 4 feet wide
   c. 1 and 2 feet wide

6 The most obvious example of the importance of timing involves
   a. poinsettias
   b. chrysanthemums
   c. leatherleaf fern

7 About two weeks after planting, the plants are
   a. graded
   b. pinched
   c. cut

8 Cut flowers are used for
   a. flower arrangements
   b. foliage
   c. hanging baskets

9 Holland is famous for its
   a. daisies
   b. roses
   c. tulips

10 Fronds are another name for
   a. buds
   b. stems
   c. leaves

Discussion

1. Why are most cut flowers grown in greenhouses?

2. What does scale mean?

3. Where do most flowers bought in the United States come from and why?

4. Why are potted plants a challenge to florists?

5. How does a pitcher plant trap insects?

6. How can temperature, rain, and wind affect plants?

7. Why is floriculture such a big business throughout the world?

8. How are cut flowers grown?

9. Why are poinsettias such a good example of timing?

10. Give some examples of exotic plants.

CHAPTER 7

Nursery Production

Student Objectives

When you have finished studying this chapter, you should be able to:

- Define a plant nursery.
- Describe materials nurseries use.
- Identify the basic greenhouse types.
- Describe what a mist system does.
- Distinguish sexual from asexual propagation.
- Describe the importance of breeding programs.

Key Terms

plant nurseries	bedding plants	propagation
stock plants	cold frames	cuttings
media	shade cloth	division

Have you ever wondered where all the flowering plants, shrubs, and trees around our houses come from? They come from plant nurseries (Figure 7–1). People who work at nurseries start with seeds or pieces of plants and grow them out to whole plants. These plants are sold in everything from 1-inch pots to 50-gallon pots. Some are even grown directly in the ground, then dug up when sold. Some of these plants must be moved with huge trucks due to their size.

Some growers produce their own plants, grow them out, then sell them. In some nurseries there are entire teams of people who only take cuttings. These people are very skilled and can take up to 8,000 cuttings in a single day.

After these cuttings are taken from the **stock plants** (plants grown for the purpose of producing cuttings), they are treated with a chemical that induces rooting, then placed in soil in greenhouses (Figure 7–2).

Many nursery businesses buy cuttings or small plants (called liners) and grow them to sell. The liners are potted up in larger containers and allowed to grow for a few months to a year. After the plants

FIGURE 7–1
The plants around your home probably originated at a nursery. *Courtesy of James Strawser, The University of Georgia.*

FIGURE 7–2
Cuttings from superior plants are used to start new plants. *Courtesy of James Strawser, The University of Georgia.*

have filled out the larger container they are either sold or transplanted into larger containers that will be sold later.

Plants bought from other sources for the grower to work with should arrive quickly and be healthy. Unrooted cuttings do not survive long. They must be handled very carefully. Frequently they are put into a cooled room until they can be processed.

Growing Media

The substances that plants are grown in are called **media**. Why not call it soil? Because many plants are not grown in soil. Over the last several years growers have begun to rely more and more on "soilless" media.

Media must combine several characteristics. They must be well drained so the roots of a plant don't rot. But at the same time, they need the ability to retain water for the plant's use.

Soilless media are usually made of a mixture of ingredients (Figure 7–3). A basic component can be peat. Peat is compressed organic matter, usually harvested from bogs that have been drained. Another common ingredient is pine bark that has

FIGURE 7–3
Soilless media is mixed from a number of ingredients. *Courtesy of James Strawser, The University of Georgia.*

FIGURE 7–4
Seeds are usually started in flats. *Courtesy of James Strawser, The University of Georgia.*

FIGURE 7–5
Bedding plants are planted and sold in nursery cell packs. *Courtesy of Getty Images.*

been crushed into small pieces. Sand is added in some instances.

Human-made, or altered, elements include vermiculite (which is actually heat-expanded rock), rock stuff, and perlite. These artificial ingredients have the advantage of being lightweight. This is especially important for nurseries that ship their plants using freight companies that charge by weight.

Containers used in the nursery industry vary widely. Many seeds are started in flats (Figure 7–4). These are trays filled with media. Usually fine, lightweight media are used to make it easier for the seedlings to grow.

Bedding plants (plants, usually annuals, that are planted in large quantities) are usually sold in cell packs (Figure 7–5). Cell packs come in several sizes, from three packs to the new larger jumbo packs. These are convenient for the person purchasing the plants. The plants are taken directly from the cells and planted.

Many plants are started in 2.5-inch pots. After growing in these pots, they are often transplanted into larger pots. Some are produced as liners to be sold to other nurseries.

For very large trees and shrubs, containers as large as 30 to 50 gallons are sometimes used. Large plants grown in the field are dug up and burlap is wrapped around the root system (Figure 7–6). This helps keep the roots moist and bound together to prevent injury. The burlap can then be planted with the tree. It decays in a short time.

Most growers use plastic pots, although a few use clay. Clay pots are much heavier than plastic ones and are easier to break. Clay pots are also porous, meaning they have tiny holes throughout them. This ensures that the plant roots will get oxygen and avoids overwatering problems. By contrast, because the clay pots are porous, water is lost faster and the plants must be watered more often (Figure 7–7).

The color of the pots is sometimes important. Black pots tend to absorb more of the sun's energy and keep the roots warm. This is good in winter, but it is sometimes a problem in summer. White pots keep the roots cooler, but they break down in the sunlight and do not last as long as black plastic pots. A few nurseries color-code their crops. Blue,

FIGURE 7–6
Large plants are dug up and the roots are wrapped in burlap. This is called balled and burlapped. *Courtesy of James Strawser, The University of Georgia.*

FIGURE 7–7
Plastic pots are lighter and will not break. Clay pots are porous and allow oxygen to get to the roots of plants. *Courtesy of James Strawser, The University of Georgia.*

pink, orange, and other colors may signify the time the plant was repotted, the type of plant, or the color of the blooms.

Growing Areas

Most nurseries have both inside and outside growing areas. Both the climate and the plants grown have an effect on the amount of outside and inside growing areas. Plants such as daylilies can be grown in fields. House plants, such as poinsettias for Christmas, must be grown in covered, heated areas.

Greenhouses are the most common enclosed growing areas. These structures allow light in and protect plants from the weather. Greenhouses are usually heated. This allows the grower to extend the growing season and produce plants all year long. Moisture can be precisely controlled on greenhouses. Plants can be given water only when they need it. This protects the plants from overwatering during very wet weather.

Most modern greenhouses are made of fiberglass or polyethylene (Figure 7–8). Fiberglass is ridged material attached to a metal or wood frame. Fairly inexpensive, fiberglass weighs less than glass, but

FIGURE 7–8
Most greenhouses are covered in fiberglass or plastic. *Courtesy of James Strawser, The University of Georgia.*

yellows with age, cutting down on the amount of sunlight that gets through. Polyethylene resembles large sheets of plastic and is even less expensive than fiberglass. However, it only lasts one to five years and then must be replaced. These large pieces of polyethylene are also attached to a metal or wooden frame.

The shape and size of the greenhouse used depends on what the nursery is producing and where it is located. In the North, the greenhouse must be able to withstand snow loads. In the South, it is more important that the greenhouse be easily cooled.

Most of the modern nurseries are putting in Quonset greenhouses (Figure 7–9). These look like a barrel cut in half. Almost all of these are built to be covered in polyethylene, although a few are manufactured for fiberglass coverings.

Freestanding even- or uneven-span greenhouses are better able to take snow loads. Some of these structures are covered with glass, but most are covered with fiberglass.

In larger nurseries, several greenhouses are connected at the sides. This is called ridge and furrow. It allows for several different climates inside while

FIGURE 7–9
Many modern greenhouses are Quonset shaped. *Courtesy of James Strawser, The University of Georgia.*

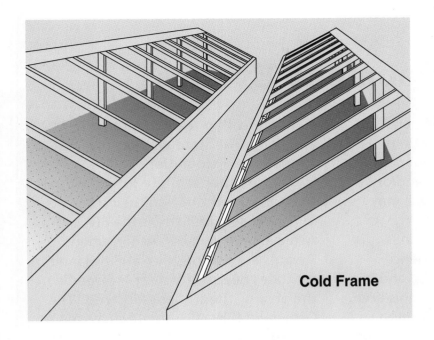

Cold Frame

FIGURE 7–10
Cold frames have no artificial heat and only provide shelter for the plants.

still being more cost efficient as far as heating and cooling goes.

A group of greenhouses at a nursery is called a range. A range can be made of all one type of greenhouse, such as Quonsets, or a mixture of types, such as Quonsets and freestanding or even-span greenhouses.

Some greenhouses are put up just to shield plants from wind and very low temperatures. These greenhouses have no supplemental heat in them. Because of this, they are called **cold frames** (Figure 7–10). Cold frames can be small structures that are mobile and just laid over plants in a bed, or they can be very large with benches just like heated greenhouses.

Outside Growing Areas

The most common outside growing structure is usually a lath house. A lath house is a metal, or more commonly, a wooden structure with **shade cloth** stretched over the top and sometimes sides. Shade cloth is loosely woven black cloth that blocks

Career Development Events

The Nursery Operations Proficiency award is an individual competition. FFA members who have job experience with turf, bedding plants, shrubs, and/or trees for production operations may apply for the Nursery Operation Proficiency. To apply, the student's supervised Agricultural Experience Program (SAEP), school classes, and FFA activities should be related to this area. Isabel was the winner of the chapter proficiency award in nursery operations. She had spent over 1,800 hours working for a local grower. Isabel's work experience included propagating shrubs, pruning field-grown trees, fertilizing container stock, and applying chemicals and fertilizer. Isabel also developed skills in communication and computer applications. She was active in her FFA fund-raising and community awareness activities and served as chapter secretary for two years. Isabel's accurate recordkeeping and wide range of experience and activities made her an excellent candidate for a proficiency award. All proficiency awards are judged on the students' application and experience. Good recordkeeping plus the development of proficiency-related skills and accomplishments are essential to become a winner. Students may

compete within their chapter, district, and state. The state winners compete for national recognition.

Another opportunity for individuals to be awarded through their interests in the nursery industry is the FFA Achievement Awards. Achievement Awards are lists of skills, abilities, and attitudes needed to enter a career in the area of the student's interest. The Ornamental Horticulture Achievement Award is divided into four areas: Career Exploration, Career Skills, Leadership Development, and Safety Practices. Each of these four areas has a list of skills and abilities, and the student has to master 80 percent of each list to receive the award. The agriculture instructor or even a worksite employer can evaluate and check the student's accomplishments.

Eddie wanted to learn more about the nursery industry and set out to win the Ornamental Horticulture Achievement Award. Eddie first worked on accomplishing the Career Exploration skills by visiting and interviewing five different horticulture operations and presenting a written report to his teacher. He also had to learn the skills required to compete as a team member in the nursery/landscape career development event and participated in the FFA chapter tree planting program. To satisfy the Career Skills, Eddie prepared a scrapbook. The scrapbook

FFA members who enjoy working with plants can get involved with projects in nurseries and apply for the Nursery Operations Proficiency Award. *Courtesy of Frank Flanders, The Agricultural Education Program, The University of Georgia.*

identified 100 plants, nursery media, pruning techniques, insect and disease problems of 20 landscape plants, and a list of tools and equipment needed in a nursery operation. Next, Eddie had to demonstrate to his teacher the proper ways to propagate a shrub, plant and prune a tree, apply fertilizer, spray insecticide, and keep records of his plants in the school greenhouse. Eddie obtained his points in Leadership Development by attending chapter meetings, participating in chapter fundraising activities, serving as chapter chaplain, and entering the public speaking contest. Eddie's employer at the landscape business checked his knowledge of safety practices. Eddie demonstrated the safe use of tools and equipment used in his job. After Eddie completed 80 percent or more of the skills as verified by the evaluator, he was presented with the official FFA Achievement Award Certificate at the school's award program.

All FFA members should strive to earn at least one Achievement Award. After deciding on your career interest, work with your parents and advisor to select the skills and experiences you want to attain.

FIGURE 7–11
Shade cloths partially block out the sun. *Courtesy of James Strawser, The University of Georgia.*

part of the sun (Figure 7–11). These black cloths are commercially available in weaves that block from 5 to 90 percent of the sunlight.

Unlike greenhouses, water is not controlled and temperature is only changed a little. These houses are used only to provide shade to those plants that do not like a lot of sun.

Some nurseries use outside beds to grow out plants. Treated boards are used to form raised beds. These are then filled with sand, soilless media, or plain soil. Plants are grown in them, then sold bare root, or without a pot. Sometimes these plants are dug up and then potted into containers for sale.

Some areas outside are covered with landscaping cloth. Landscaping cloth is a thick, black cloth that forms a barrier to weeds and holds down erosion. Containers are put directly on the landscaping cloth and grown out, then sold.

Another area that is used for growing plants outside is the ground itself. Plants are put into the ground, allowed to reach a good size, then harvested and sold bare root.

Mist Areas

Some plant reproduction techniques involve the use of mist. Tiny droplets of water from irrigation lines keep the plants moist so they don't use up their water reserves (Figure 7–12). There are two main types of mist systems used. The first type is a timed system. A solenoid wired into the irrigation lines controls the flow of water. The solenoid is controlled by a timer. The water goes on for a short time (a few seconds) every five minutes to every hour, depending on circumstances.

The second type uses a mechanical leaf. A piece of wire is attached to an electrical device. When the screen (the mechanical leaf) is wet, the weight of the water on the screen tilts down, trips the solenoid,

FIGURE 7–12
Irrigation systems in modern green-houses are timed to water at certain times. *Courtesy of James Strawser, The University of Georgia.*

and shuts off the water. As the screen dries out, it becomes lighter. When this happens, it tilts up and trips the solenoid to let water through. The water makes the mechanical leaf heavy again and turns the solenoid off.

Propagation

Plant reproduction, or making several plants from one plant, is called **propagation.** This is done several ways, depending on the plant and the nursery operation.

Some operations produce their own seed and grow plants from them. Others buy seed to begin plant production. Some only handle pieces of plants. Others may combine the two main propagation methods.

Sexual Propagation

Sexual propagation refers to producing plants from seeds. Plants have been producing seeds for many millions of years. Humans began taking advantage of this at the dawn of civilization. Plants that they found useful were propagated to produce new plants.

Sexual propagation begins with the flower. Flowers differ in size, color, and parts. Some plants have only female flowers on them, others only male. Other types of plants have male and female flowers separately on the same plant. Most plants, however, produce flowers that are both male and female.

The female parts of a flower, called the pistil, consist of the stigma, the style, and the ovary. Egg cells are produced in the ovary.

The male parts of the flower, called the stamen, contain the anther and filament. The anther produces pollen, or sperm cells. The pollen, carried by the wind or insects, lands on the stigma. This is called pollination. It then forms a pollen tube through the style to the ovary. When the sperm cell actually unites with the egg cell fertilization has occurred.

Flowers fertilize each other or themselves to produce seeds. When they fertilize themselves they are self-pollinating. When they pollinate another flower they are cross-pollinating. When the flowers pollinate the male and female cells are united. This forms the zygote that develops into the seed.

Many nurseries that produce annual bedding plants start with seed they order from another nursery (Figure 7–13). They germinate the seed, then grow it out. Germination is when a seed starts to grow into a plant.

Growing plants from seed is less expensive than growing plants from cuttings. It enables a grower to produce many thousands of plants quickly.

Asexual Propagation

Sometimes it is less desirable to start plants from seed, especially for large trees and shrubs. It takes longer to get a salable plant starting from seed. In some instances, a plant doesn't "come true" from

FIGURE 7–13
Annual plants are often started from seed. *Courtesy of James Strawser, The University of Georgia.*

seed. That means that the plant from a seed may not be exactly like the parent plant. This is because of cross-pollination where characteristics are obtained from both parent plants.

For this reason, many growers produce their plants using asexual propagation. This method starts with a part of a plant itself instead of a seed produced by the plant.

Cuttings

The most popular form of asexual propagation is taking **cuttings** from the plant. A piece of the stem (or sometimes a leaf) is taken from the plant. It is usually treated with a rooting hormone (a chemical that makes the plant form roots), then placed in media. The cutting is then placed under mist until it has formed roots and no longer needs the extra water (Figure 7–14).

Propagating plants this way assures the grower of a crop that will be almost exactly like the original. In addition, the plants will be uniform in growth. This is especially important for landscapers who buy many plants for hedges or beds.

Some plants cannot be propagated this simply. In this case, other asexual methods are used. Some

FIGURE 7–14
Root hormones are applied to help a cutting grow new roots. *Courtesy of James Strawser, The University of Georgia.*

FIGURE 7–15
Many plants can be propagated by separating or dividing the roots or bulbs. *Courtesy of James Strawser, The University of Georgia.*

FIGURE 7–16
Holly is valued for its bright leaves and berries. *Courtesy of James Strawser, The University of Georgia.*

plants are propagated by air-layering. The stem is peeled all the way around, a rooting hormone is painted on the wound, and the area is then wrapped in damp media, covered with plastic, and allowed to form roots while still getting water from the parent plant.

Other plants must be propagated by root cuttings. The roots are cut off much as the stems are, then allowed to produce a plant.

Tissue Culture

Micropropagation, or tissue culture, produces an entire plant from a single cell or group of cells of the parent plant. Although it takes quite a bit longer than producing plants from cuttings, it has the potential to produce millions of plants from a single branch of a plant.

Still other plants must rely on **division** or separation for propagation (Figure 7–15). This involves digging up part of a plant just below the soil surface and separating it into several parts. Some plants can be gently pulled apart, called separating. Other plants must be cut apart into sections; this is called dividing. These parts are then allowed to grow into whole plants.

Plant Production

Let's follow a typical plant from establishment to selling. One of the most popular plants sold in nurseries today is the holly. Hollies are valued for their glossy green foliage, red berries, and screening ability (Figure 7–16). A row of hollies can screen a busy highway or noisy neighbors. They are common in landscaping plans.

Propagating *Ilex crenata*, the Japanese holly, begins with stock plants. These are large plants that are either planted on the nursery grounds to be used as cutting material, or older plants in large containers.

② The cuttings must be taken from the right part of the plant. Branches that look sick or were heavily shaded by the rest of the plant should be avoided. Old branches with thick bark should also be left. ③ Very young, weak tips may rot under mist. Usually young, firm wood is taken. The cuttings are between 4 and 6 inches long.

The best time to take cuttings is in the cool morning. If exposed to heat and direct sunlight, the cuttings may suffer enough damage to be unusable. Cuttings are kept in a cool, shaded place and kept moist until ready to process. Depending on the operation, a few hundred cuttings may be taken or several thousand before they are all processed.

④ The lower leaves are taken off to expose the bottom node. Often the bark is scraped off in a small area to help the plant to form roots. A powdered or liquid rooting hormone is then applied to the bottom of the stem.

⑤ The cuttings are then stuck into moist media in small pots, bare root trays, or outside beds under mist. Keeping the new cuttings moist is especially important. If allowed to dry out too much, the cuttings will die. If the area is kept too wet, fungus and bacteria may become a problem.

⑥ The cuttings stay there until they form roots. They are misted less and less frequently. After three to four weeks, the cuttings are weaned off the mist.

⑦ Once the root system has filled the small pots, they are repotted into 1-gallon containers or sold as liners. When the roots have filled the 1-gallon pots they are either sold or repotted again. Hollies are available in as large as 50-gallon pots. This size is rather expensive, however, and the majority of hollies are sold in 1- to 3-gallon pots (Figure 7–17).

When growing out, the hollies are fertilized, either with constant liquid fertilizer, or a granular fertilizer put on one to three times a year. They are sprayed for pests, and the pots kept weed free.

FIGURE 7–17
Young holly plants are sold in pots.
Courtesy of James Strawser, The University of Georgia.

Water is used when needed, usually twice a week or so, depending on weather conditions.

Summary

There are many things to consider when thinking about the nursery business. Nurseries vary widely in the plants they grow and the methods they use. The nursery business is complex and fascinating.

CHAPTER REVIEW

Student Learning Activities

1. Take an inventory of the plants around your home or school. Determine which of these originated in a nursery.

2. Visit a nursery in your area. Find out which plants sell the most. Think of some reasons why these particular plants are the most popular. Report to the class.

3. With your teacher's assistance, take cuttings from a plant and root them. Grow the plant in a pot and at the proper time plant it at your home.

4. Examine the plants around your school. Identify each and determine how each was propagated.

True/False

1. ___ Plants are the basis of the nursery industry.

2. ___ Unrooted cuttings survive a long time.

3. ___ Vermiculite is a human-made element.

4. ___ The most common outside growing structure is a greenhouse.

5. ___ Landscaping cloth is used solely to keep insects away from plants.

6 ___ The most popular form of asexual propagation is taking cuttings from plants.

7 ___ Micropropagation produces an entire plant from a single cell of the parent plant.

8 ___ The best time to take cuttings is in the afternoon.

9 ___ Stock plants are grown for the sole purpose of producing cuttings.

10 ___ All substances that plants are grown in are called media.

Multiple Choice

1 Human-made elements include
   a. peat
   b. bark
   c. perlite

2 The most common enclosed growing area is the
   a. nursery
   b. greenhouse
   c. lath house

3 Most modern greenhouses are made of
   a. plastic
   b. metal or wood
   c. fiberglass or polyethylene

4 Greenhouses put up to shield plants from wind and temperatures are called
   a. cold frames
   b. ranges
   c. lath houses

5 Shade cloth is used to
   a. kill weeds
   b. block out the sun
   c. hold down erosion

6 Sexual propagation refers to producing plants from
   a. seeds
   b. leaves
   c. cuttings

7 Egg cells are produced in the
    a. ovary
    b. pollen tube
    c. stigma

8 When flowers pollinate another flower it is called
    a. self-pollination
    b. cross-pollination
    c. fertilization

9 Freestanding even- or uneven-span greenhouses are better able to take
    a. water loads
    b. soil loads
    c. snow loads

10 The most common outside growing structure is the
    a. greenhouse
    b. lath house
    c. range

Discussion

1 Why are plastic pots more common than clay pots?

2 Why is the color of plant pots important?

3 What are the two main types of mist systems?

4 What is the difference between sexual and asexual propagation?

5 List the male and female parts of a flower.

6 What is air-layering?

7 What are media?

8 What is a lath house?

9 What is the difference between self-pollination and cross-pollination?

10 When is the best time to take cuttings? Why?

CHAPTER 8

Landscaping

Student Objectives

When you have finished studying this chapter, you should be able to:

- Define the term landscape.
- Describe the duties of a landscape architect and a landscape designer.
- Describe the different types of plant materials most commonly used by landscape architects and designers.
- List and describe the phases of landscape development and construction.
- List and describe the five principles of landscape design.
- Describe the duties of a landscape contractor.
- Define the term interiorscaping.

Key Terms

landscaping	emphasis (focal point)	variety and unity
landscape architect	balance	topiary
landscape designer	proportion and scale	interiorscape
landscape contractor	rhythm	

Designing the Landscape

What do you think of when you hear or read the word "landscape?" Most people imagine mountains, water, forests, or wild meadows. The term landscape actually means a view of any natural scenery as seen from one perspective (Figure 8–1). **Landscaping** is the act of taking a certain plot of ground, usually in its natural state, and redesigning it to be more comfortable, attractive, and useful for humans or other animals. For example, your local park has most likely been landscaped.

If you have a backyard, it probably has been landscaped, too. Even the habitat areas in zoos are a result of careful planning so that the animals feel that they are in their natural homes. Some people like the look of natural areas such as forests or wild meadows, whereas others prefer to arrange landscapes to better suit their needs and preferences. They may do this by planning and creatively placing different plant materials on their land. The choice of most Americans to do this has developed into the multi-billion dollar industry of landscaping (Figure 8–2). This business employs thousands of people in many different areas of landscape development.

FIGURE 8–1
The term landscape actually means a view of any natural scenery as seen from one perspective.
Courtesy of Valerie Mosley.

FIGURE 8–2
Landscaping has developed into a multi-billion dollar industry. *Courtesy of Valerie Mosley.*

Landscape architects or **landscape designers** usually create landscapes. Both professions are trained in basic horticulture and design. They also work extensively with plants and other outdoor elements to create landscapes. However, the landscape architect differs slightly from the designer. The landscape architect is trained to design not only where roads, walkways, fountains, water drainage systems, and small outdoor structures will go but also how they will be built. The most important skills of both professions are the ability to work well with other people, listen, and communicate their ideas through drawing, speaking, and writing (Figure 8–3).

Landscape contractors are the people who actually construct the landscape plans. They assemble and install all of the materials exactly as the landscape architect or designer has instructed in the plan. Landscape contractors must also be able to work well with other people and communicate, as they will often be in contact with the client or the landscape architect/designer. The contractor usually employs a team of landscape technicians to

FIGURE 8–3
Landscape architects and designers must be able to work well with others, listen, and communicate their ideas through speaking, drawing, and writing. *Courtesy of Getty Images.*

Landscape Contractor

Operate heavy equipment
- Bulldozer/backhoe/dumptrucks/
  crane
- Spread gravel/sand/dirt
- Masonry — building stone/brick
  hello/stone
- Plumbing to install pond/waterfall
- Install Trees/Shrubs/gardens,
  etc.

Landscape Architect — draws up
the paper plans

Landscape Contractor takes
the paper plan and makes
it happen by installing
planting what is prescribed
in the paper plan.

help install the design. Landscape contracting companies also have maintenance crews to help keep the landscape in good condition once installation is complete. Sometimes landscape architects and designers will form a landscape contracting company as well. In this way, the company not only designs the plans but also can install the new landscape.

Plant Materials

Landscape design is a complex art. Like many artists, the landscape architect or designer brings together various elements to create an attractive, functioning, and safe place. Several different materials are used in landscaping such as rocks, water, and human-made features such as walkways, decks, and arbors. However, in almost all schemes, plants are essential to landscape designs (Figure 8–4).

Plants are living organisms that have distinct shapes, habits, and needs. They also change as they grow. Many different types of plants may be used in landscape designs. Those most commonly used are called woody landscape plants and herbaceous ornamental plants.

Woody landscape plants are the foundations of most planting designs. The term *woody* indicates that the plant's stems and branches contain wood tissue. Most trees and shrubs are woody landscape plants. However, this category may be divided into two types of plants: deciduous and evergreen.

Deciduous trees and shrubs are woody landscape plants that completely shed their leaves each year during the fall or early winter. Before separating from the tree or shrub, the leaves begin a dying process in the fall. At this time, the leaf color may change from green to bright red, yellow, orange, or brown depending on the type of tree. Some examples include oak and maple trees.

FIGURE 8–4
Several different materials are used in landscaping such as rocks, water, and hardscape features. However, in almost all schemes, plants are essential to landscape designs. *Courtesy of Valerie Mosley.*

Evergreen trees and shrubs shed some of their leaves occasionally throughout the year; however, a great majority of the leaves remain on the plant. As the name says, evergreens are always green. This adds important color during the winter when most grasses have turned gray and other plants have lost their leaves. Although they do not offer the brilliant fall colors of deciduous plants, the evergreens can be effective year-round for screening unsightly views and breaking strong winds that blow through the site. Pine trees and holly shrubs are examples of evergreens.

Herbaceous ornamental plants are also frequently used in landscape designs. The stems and branches of herbaceous ornamentals are not woody. These plants usually die down at the end of the growing season, and therefore do not provide year-round interest in the landscape. However, during their growing season they may offer incredible blooms and color. These flowers not only add strong visual interest for a garden's visitors but also attract wildlife such as birds, butterflies, and bees. Three types of herbaceous ornamental plants are annuals, biennials, and perennials (Figure 8–5).

Annual plants live only for one growing season. During the spring and summer they will live and produce leaves and blooms. At the end of the growing season the plant will die and will not return. Biennials are plants that will live for two years only, dying after the second growing season. Perennials will return for every growing season.

FIGURE 8–5
Annual and perennial plants are sometimes used in designing a landscape. *Courtesy of Getty Images.*

Phases of Landscape Development

There are four basic phases to landscape development:

1. Site inventory and analysis
2. Landscape design

3. Landscape construction
4. Maintenance

Each step is important and should be conducted in the order as shown.

Site Inventory and Analysis

The landscape architect or designer will first meet with the client, or project owner, to understand what they want and need in their project. A landscape project can be a variety of sizes. It could be a big project, such as a shopping center or a school, or it could be a smaller project, such as a backyard or a park. After this meeting, the designer begins the first step in landscape development: site inventory and analysis. This process involves visiting the site to identify what is currently there, such as any existing structures, trees, or other vegetation. It also should include understanding the wind directions, the slope (whether the site is flat or hilly), the soil type, and the way water flows over the project site.

Once the designer has inventoried what is there, the information will be analyzed. Certain questions should be asked at this stage, depending on the project. For example, what, if any, of the existing vegetation should be removed? What types of plants grow best in this kind of soil? Where is true south (plants placed on the south side of the site will receive the most sun)?

Every plot of ground is unique. Performing a site analysis helps the designer understand the character of the project site. This will ensure a more successful landscape design.

Landscape Design

When the designer has completed the site inventory and analysis, he or she moves to the landscape design phase. In this phase, the designer pulls together all of the information on the site with the needs of the client and forms a landscape design. The creation of the design takes a lot of time and goes through a long process of designing, evaluating, then redesigning and developing the plans.

The Design Principles   There are five basic design principles that should be considered in designing landscapes: **emphasis (focal point), balance, proportion and scale**, **rhythm**, and **variety and unity**.

All of the principles are essential components. They must work together to create a strong, successful design. The ideas and artistic ability of the landscape architect or designer determine how each principle is used.

Emphasis is also called the focal point of the design. Every design should have some element that stands out enough to draw attention. Focal points communicate how people should move throughout the landscape. They also can show how the landscape should be viewed. A focal point can be an unusual plant, a statue, a pool or fountain, or a spectacular view. Sometimes a focal point can be a building (Figure 8–6).

FIGURE 8–6
A focal point can lead visitors through the garden. *Courtesy of Neal Weatherly, The University of Georgia.*

Although the point of emphasis should be strong enough to get attention, it should not be too strong. Otherwise, it overpowers the entire landscape design. Placing too many focal points will confuse people. The purpose of the focal point is to order the design by giving the eye a comfortable place to rest.

A balanced landscape is a comfortable landscape. This means that one part of the landscape design does not appear to weigh more than another area. There are two types of balance: symmetrical and asymmetrical. Symmetrical balance is the more formal type.

Both types of visual balance are used in landscape designs. If there is a really large, dark plant on one end of a building, the designer can locate another of the exact same plant in the same place on the opposite end of the building to balance it. This would be symmetrical balance (Figure 8–7). Using the same first plant, the designer could instead locate a group of about three smaller, lighter plants of a different type somewhere on the opposite end of the building. This contrast would create asymmetrical balance (Figure 8–8).

FIGURE 8–7
This is an example of a symmetrically balanced residential landscape. *Courtesy of James Strawser, The University of Georgia.*

FIGURE 8–8
This landscape is asymmetrically balanced. *Courtesy of W. T. Smith.*

Proportion means that the plant material and human-made structures should generally be in scale with each other. For example, a very large house should have large trees and shrubs nearest the house. It would look uncomfortable and out of scale if its landscape had only small trees and groundcovers. Of course, people are different. Some may prefer using plants that are not in proportion to the building and other plants. For the most part, designing a well-proportioned landscape will create a more comfortable area (Figure 8–9).

Think of your favorite song. All songs have a particular rhythm that helps us remember them.

FIGURE 8–9
These low-growing plants are in good proportion. They enhance the gentle slope of the ground. *Courtesy of Terry Hamlin, Georgia Department of Education.*

Unity & Variety

A landscape works the same way. Just as musicians repeat chords in their songs, the landscape designer can place the same type of plant in different places to establish a rhythm. For example, azaleas placed in certain places throughout the garden will give a sense of continuity. This brings us to the last principle of design: unity and variety.

Unity can be a very difficult principle to master, but it is extremely important. A brick or stone walkway, also called hardscape, can be unifying as it ambles from one area of the landscape to another. Establishing a rhythm through plants or hardscape materials will help to unify a design. All the elements of a landscape design should work together as a whole, and the different parts should not compete for attention (Figure 8–10). However, landscape architects or designers usually group variety with unity as the final design principle. This is because the two should check each other. When there is too much variety, unity is lost. When there is too much unity, the design becomes boring. Ideally, there should be just enough variety to keep the design appealing, yet enough unity so that the entire design works together.

FIGURE 8–10
A landscape design should work together as a whole. It should not have several very different parts that compete for attention. *Courtesy of Getty Images.*

The Landscape Plan   After completing the design process, the landscape architect or designer will create a landscape plan to communicate his or her ideas to the client. A plan is any drawing that represents the site or subject area as if viewed from above (Figure 8–11). Imagine that you are hovering over your desk. If you were to draw what your desk looked like from that perspective, you would have created a plan view of your desk. It is the same with a landscape plan. The landscape plan shows how the site has been designed from a plan view.

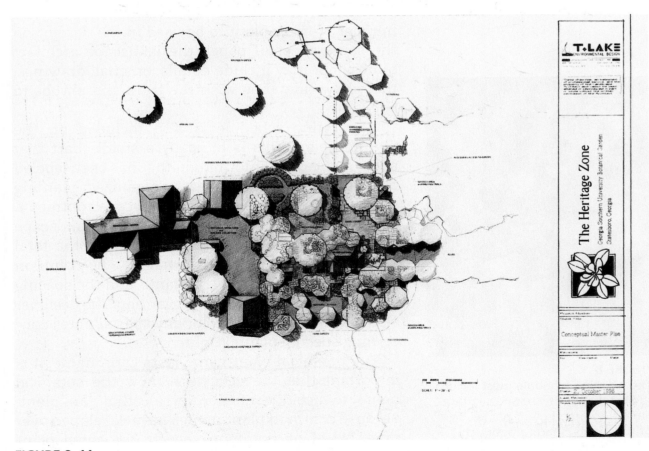

FIGURE 8–11
A landscape plan is a drawing that represents the site or subject area as if viewed from above.
Courtesy of T.LAKE Environmental Design.

On this particular plan, the designer will identify walkways, roads, benches, water fountains, plant masses, or whatever has been included in the design. Landscape plans are usually colored, or rendered, to make them more understandable and attractive to the client.

The Planting Plan Along with the landscape plan, the landscape architect or designer creates a planting plan. The landscape contractor uses this plan to lay out and install the landscape plan, so it must clearly communicate the designer's ideas. The planting plan is a simple drawing without color that focuses specifically on the plants in the design. This plan must indicate the placement, name, and number of every plant to be used in the design. On the same sheet of paper, the landscape architect or designer may include details, or small drawings, that show how they wish the trees and shrubs to be planted.

A planting schedule is also included on the planting plan. This is basically a small chart that adds up the plant information that has been labeled on the plan. A good, clearly organized planting schedule is very helpful to the landscape contractor. It will indicate all that is necessary for landscape construction. The schedule must include the total quantity, the botanical plant names, the common plant names, the size of the plants, and the spacing between plants at the time of planting. The designer also may include an area for any special notes concerning specific plants (Figure 8–12).

The botanical plant name is its Latin name. It is important that the designers know the botanical name to avoid confusion in locating the plant. Because common plant names have developed over time, based on common speech, the same plant may be called by different names in different areas. For example, the beautiful small tree with the botanical name *Chionanthus virginicus* is commonly

FIGURE 8–12
The planting schedule must include the total quantity, botanical plant names, common plant names, size of the plants, spacing between plants at the time of planting, and any important notes about a particular plant. *Courtesy of Getty Images.*

called a Fringe Tree in the southern United States but a Grancy Grey Beard in the north. A sample planting schedule follows.

QTY	Botanical Name	Common Name	Size	Spacing	Notes
20	*Chionanthus virginicus*	Fringe Tree	15 gal.	20' on center	Multitrunk
18	*Itea virginica*	Virginia Sweetspire	5 gal.	6' on center	
35	*Raphiolepsis indica*	India Hawthorn	3 gal.	3' on center	

Landscape Construction

After the client has approved the landscape design and the plans have been completed, it is time to begin construction (Figure 8–13). Spring or fall is the best time to construct landscapes. Planting at these times gives the new plants time to grow and adjust to their new home before facing the drying heat of summer or the strong, cold winds of winter. If an irrigation system is to be added, it will be installed before any planting begins. The landscape contractor also will build any hardscape materials for the

FIGURE 8–13
After the plan is completed and approved, a contractor will construct the landscape. *Courtesy of Terry Hamlin, Georgia Department of Education.*

Career Development Events

The FFA Nursery/Landscape Career Development Event is a competition for teams consisting of four members. Each member competes individually and the final team score is the sum of the three highest individual scores. The contest consists of four parts: (1) identification of plant materials; (2) placing classes, problem solving, and equipment; (3) general knowledge; and (4) practium.

Identification of plant materials consists of plants selected from the State Nursery/Landscape list and displayed for visual identification. Students may not touch the live plant material but can use a magnifying glass. The placing classes, problem solving, and equipment phase of the contest consist of five items with any combination of judging classes, problem solving, and questions about landscape/nursery equipment. The judging classes are plant materials or landscape plans in which the student places the materials from best to worst. Plant material is placed based on its structural form, market value, and health. The landscape plans should be placed by the criteria of design principles, plant selection and placement, and use of basic drafting methods. Nursery/landscape plants will be selected from the identifi-cation list and will be the same species and cultivar in each class. Plants are placed by structure, market value, and health.

The general knowledge part of the nursery/landscape event consists of 50 multiple choice questions. This phase of the event tests for knowledge and understanding of all areas of the nursery/landscape industry. Areas for testing include plant materials, plant disorders, cultural practices, landscape design and construction, supplies and equipment, safety, equipment operation and maintenance, interpersonal relations, salesmanship, and records and reports. In the practium, students identify plant disorders and answer questions about a given landscape plan. Plant disorders include diseases, insects, weeds, and nutritional disorders. Disorders are displayed for the student to view with the naked eye or magnifying glass. The student is asked to identify the problem and those pesticides or other treatments necessary to remedy the problem. The landscape plan practium tests the student's ability to read a landscape plan and knowledge of landscape design principles by answering multiple choice questions.

After all team members have completed each phase of the event, each individual score is calculated. The top three individual scores are added to yield the total team score. If there is a tie between two team scores, the written exam score is used as a tiebreaker for first and second place teams. State winning teams advance to the National Nursery/Landscape Event.

As part of the Nursery/Landscape Career Development Event, students must determine where plants are to be used in landscaping. *Courtesy of Carol Duval National FFA Organization.*

The Turf and Landscape Management proficiency award is an individual competition. This proficiency award can be applied for if an FFA member's job experience has focused on the installation, maintenance, and design of landscape plants and materials. Vincent is a senior and the winner of the chapter proficiency for turf and landscape. He started his own lawn care service when he was in the eighth grade. He now has contracts with several local businesses and many homeowners. Skills such as mowing, aerating, dethatching, fertilizing, and weed control are part of Vincent's business. Last year he made over $4,000 before his expenses were subtracted. Vincent was on the FFA senior high parliamentary procedure team and livestock show team, and served as chapter vice president. All proficiency awards are judged on the student's application and experience. Good record keeping of inventories and expenses plus the development of proficiency-related skills and accomplishments are essential to be a state and national winner.

project, such as walls or walkways, before beginning to install the plants.

A copy of the landscape plan and the planting plan will be given to the landscape contractor whom the client has selected. The contractor will usually consult with the landscape architect or designer if there are any questions, and will order all of the materials necessary to construct the landscape. Then, either the landscape architect, designer, or contractor will begin laying out the design according to the plan.

Once the design has been laid out on the project site, the contractor's team will begin preparing the soil for planting. When the soil is tilled and ready, installation begins. At the time of planting, some trees are so large that they require bulky equipment such as tractors or cranes to place them. It is a good idea to first locate and plant the largest materials, usually trees, to avoid damaging other plants. Next, the shrubs can be planted. The contractor finishes with smaller plants and the installation of sod or grass seed for the lawn.

Mulch is the key to healthy planting beds. After all planting is complete, the landscape team will mulch the planted areas with pine straw or wood chips at about 2 to 3 inches deep. This will insulate the plants during the winter months. It also will help keep the soil moist during the hot, dry summer months.

Landscape Maintenance

Once landscapes have been installed, they must be maintained to keep them healthy and attractive. Often a landscaping company or an individual offers landscape maintenance. This means that at certain times the company or person will mow the lawn, weed the beds, ensure irrigation systems are working properly, and check the plants for pest problems.

Landscape maintenance begins in the early spring and continues until late fall. It begins with a general cleanup followed by evaluation. This involves weeding and edging beds as necessary, removing dead leaves and branches from trees and shrubs, mowing the grass, and adding fresh mulch to the planting beds. The landscape maintenance person or technician uses a variety of tools during the cleanup. For small areas, simple hand tools and lawn mowers may be used. For the larger lawns of businesses or parks, bigger equipment is used such as motorized leaf blowers and tractors with a mowing blade attachment. A general evaluation is then done. The technician looks for pest damage on all of the plants, including the lawn.

Lawn Care Trees, shrubs, and herbaceous plants can require attention such as pruning, irrigation, and fertilization. But, probably the most time and money spent in landscape maintenance is on lawn care. Imagine large parks or residential landscapes that you have seen. Many of them likely have big, smooth, green areas of lawn. Although the lush lawns look simple, they don't grow and stay that way by themselves. Sometimes when a lawn is suffering from pest problems, large areas of the grass die out. A chemical treatment is usually needed if there is a serious pest problem (Figure 8–14). This must be done by licensed applicators. These people know the proper pesticide to use and the correct amounts. They must follow strict guidelines to prevent problems with the toxic materials.

Fertilizing is the next step. Grasses that grow in the cool part of the year receive their first fertilization in early spring. Those growing best during warmer weather are fertilized in late spring. Many people prefer that their lawns stay only one type of grass. When this is desired a chemical weed killer is used to keep the other grasses and wildflowers out

FIGURE 8–14
Plants sometimes have to be treated with chemicals to control pests. *Courtesy of Terry Hamlin, Georgia Department of Education.*

FIGURE 8–15
Irrigation systems provide water for the landscape plants. *Courtesy of James Strawser, The University of Georgia.*

of the lawn. This weed killer is generally applied in combination with the first lawn fertilization.

Grasses also must get enough water to do their best. Water is usually added by irrigation. This can be done with overhead irrigation such as moveable sprinkler heads. Underground irrigation also can be used. This is when the water lines of an irrigation system are buried and only the sprinkler heads pop up when the water is turned on. Irrigation is especially important in the early growing season (Figure 8–15). If grass does not get the water it requires when it first starts growing, it may remain stunted for the rest of the year.

Mowing usually requires the most time spent on a well-maintained lawn. If the grass is healthy and lush, it will grow quickly. Many of you are probably familiar with a lawn mower. You may even mow the lawn weekly and can understand the amount of time mowing requires. To remain healthy, most grass needs to be mowed at 2½ to 4 inches in height (Figure 8–16). This depends on the type of grass and the height with which the homeowner is comfortable. In most cases, such details are left for the maintenance person to decide.

FIGURE 8–16
For grass to remain healthy and attractive, it has to be mowed at the proper height. *Courtesy of James Strawser, The University of Georgia.*

With proper fertilization, pest control, irrigation, and mowing, a lawn can usually be kept beautiful for many years. Lawn areas can provide wonderful areas for playing, running, or relaxing outside. A lawn is much less expensive to install than areas planted with trees, shrubs, or groundcovers. However, lawns also can be time-consuming and expensive to maintain. For this reason, many home-owners are now requesting landscape designs with smaller lawn areas and larger naturalized areas, mulched areas that are filled with plants that grow there naturally.

Pruning Pruning is the removal of selected branches of shrubs or trees. This is done to maintain the size or shape of a plant (Figure 8–17). Sometimes part of the plant is damaged. This damaged portion must be removed. Afterward, the plant may need to be pruned into a more pleasing shape.

Pruning can be considered an art form. Many gardens in Europe have plants that are pruned into topiaries. **Topiary** can be anything from simple geometric forms to intricate animal shapes (Figure 8–18). Many painstaking hours go into these works of art. As the plants grow, they are pruned to

FIGURE 8–17
Pruning is the removal of some of the branches. This helps to shape the tree or shrub. *Courtesy of James Strawser, The University of Georgia.*

allow certain parts to grow more than others. This gives the topiary its shape.

In landscape maintenance, pruning is one of the most important procedures to do correctly. A plant that is pruned badly might not recover, or it could actually die. Spring is the best time to prune plants; however, some also can be pruned in the winter.

Most plants that bloom in the spring should be pruned immediately after they have bloomed and not in the fall. Fall pruning cuts off buds that

FIGURE 8–18
Topiary can be anything from a simple geometric form to an intricate animal shape. *Courtesy of Frank Flanders, The University of Georgia.*

would normally bloom the following spring. Another problem with fall pruning is that pruning often stimulates a plant to produce new growth. This new growth is more susceptible to cold damage during the coming winter.

It is difficult to see the branching habit of deciduous plants in the summer. This makes it harder to decide correctly which branches to prune. The plants are also more susceptible to insect and disease damage and drought stress at this time of year. Therefore, most professional pruning is done during the late winter or early spring.

During the winter, pruners can see exactly what they are pruning. Also plants are in their dormant state in the winter months. This means that they are not actively growing above ground though their roots and branches remain alive. Less damage is done to the plant when it is pruned in its dormant state.

Interiorscaping

Plants not only purify air and produce oxygen necessary to humans but also help create comfortable places for living, working, and shopping. As more and more people recognize the benefits of plants in human environments, landscape designs are being created for the indoors (Figure 8–19). Think about a mall or other place of business you have been to recently. Chances are there was some type of indoor landscape or **interiorscape** inside the area. An example of interiorscaping may include big potted plants, large bed areas, or gushing water fountains.

The development process is the same for interiorscapes as for outdoor landscapes. However, there are some unique restrictions. The designer often does not have the freedom of expanding or moving planting beds. The height of the building also limits the size of the plants used.

FIGURE 8–19
Interiorscapes bring the concepts of landscaping indoors. *Courtesy of James Strawser, The University of Georgia.*

Obviously, plants do not grow naturally inside. Light, soil, and water—everything necessary to a plant's health—must be provided. In the past, most buildings were not built with the health of plants in mind, and had to be modified to support plant life. Today, architects are beginning to design buildings that use natural light and provide areas for plantings. This makes it easier to include interiorscapes, although rarely is the lighting the same as that outside. In this instance, supplemental lighting must be provided. This means there must be a source of electricity as well as a place to anchor the lighting fixtures.

The choice of plants that can stay healthy inside buildings is limited. Most are tropical plants. These are plants that usually do well in extreme shade and can take warmer temperatures year-round. Fruit trees and most flowering shrubs cannot be grown inside because they must have a certain amount of cold before they will bloom.

Interior plants must be able to take a lot of abuse. Often the plants must tolerate indoor air pollution. Dust builds up on their leaves quickly in this environment. As a result, photosynthesis is reduced and the plants look bad. The plants must also have a reliable water source nearby. The architecture of the building or the plant containers must be designed to hold water seepage. Clients may get upset if they have water stains on their carpets. These factors must be considered as the plants are placed in each location.

Just like their outdoor components, indoor plants often need fertilizer. They do require less, however, and it should be applied less frequently. Fertilizing schedules have to be developed and implemented to optimize the beauty of the plants.

Insects and other pests can damage indoor plants. It is especially important to spot these

FIGURE 8–20
Interiorscapes require a lot of maintenance. *Courtesy of James Strawser, The University of Georgia.*

problems before they become serious. Because of the closed environment in which they grow, pest problems can spread rapidly. Because of the high rate of foot traffic around the plants, the pesticides must be used carefully when a serious pest problem does occur.

When properly planned and maintained, indoor plantings can give a business, home, or office a healthier and more comfortable atmosphere (Figure 8–20).

Summary

Landscaping is a complex and fascinating field that extends in many directions and requires diverse skills. Landscape architects, designers, contractors, technicians—all are crucial to the success of any landscape design whether indoors or outdoors. Although hard work is necessary in all of these occupations, it is rewarding to see a beautiful job well done that will be enjoyed for many years.

CHAPTER REVIEW

Student Learning Activities

❶ Art, like landscapes, should contain the five elements of design. Go to the library and find a photograph, picture, or painting that you like. Try to identify the design elements in your picture. You may want to lay pieces of trace paper over the image and draw the different elements of your picture on the trace—that means you should have five different pieces of trace with one element sketched on each. When these sheets are put together, you should have a rough sketch of the original image. Explain your image and trace sheets to your classmates.

❷ Locate what you consider to be the most beautiful landscape in your area. It can be a park, a home landscape, a natural area—whatever you like most. Draw a sketch or take a photograph of the area. Why do you like it? Are the design elements present in the landscape? You may wish to use the trace overlay technique from exercise 1 to identify them. Share your findings and opinions with the class.

❸ Closely examine the landscape around your home. If you do not have a front- or backyard, examine a park, your school's grounds, or a neighbor's yard. What do you like or dislike about it? Why? Can you identify any design principles?

❹ Make a sketch of your school building and its surrounding grounds. How would you design the area? Think about the uses of the landscape space and who will be maintaining it. Create a landscape design for the area. In your sketch, identify the types of areas (play area, vegetable garden, wildlife areas, etc.) and the types of plants that you think should be in the landscape. You do not have to know the specific plant names, but you should label what kind of plants you want. For example you could label a tree as being small, deciduous, and flowering with red fall color.

True/False

1 ___ The focal point of a design also can be called the point of emphasis.

2 ___ Perennial plants live for only one growing season and never return.

3 ___ The leaves of evergreen plants change color in the fall.

4 ___ The stems and branches of herbaceous ornamental plants contain wood tissue.

5 ___ Before beginning the design process, the landscape architect or designer should visit the project site to inventory and analyze its existing conditions.

6 ___ Mulching plants will insulate them from the cold and will protect them from drying out in the heat.

7 ___ Fall is the best time to prune plants.

8 ___ Fruit trees and flowering shrubs make excellent indoor plants.

9 ___ A planting plan is a simple drawing with labels, details of how the materials should be planted, and a planting schedule.

10 ___ Designers and contractors should always use the common names of plants to avoid confusion in finding them.

Multiple Choice

1 Before beginning the landscape design phase, it is very important to complete a
    a. landscape plan
    b. site review
    c. site analysis
    d. design element test

2 Identify the item that does not belong in the list below. A planting plan must include the
    a. Latin and common plant names
    b. number of each plant to be used
    c. size of the plants at time of planting
    d. name of the nursery that will provide the plants

3 A _____ creates landscape designs including planting plans and hardscape plans, but also may design where and how roads, outdoor structures, fountains, and water drainage systems will be built.
   a. landscape designer
   b. landscape architect
   c. architect
   d. landscape gardener

4 Once the landscape design and plans are complete, they will be given to a _____, who will begin construction of the design.
   a. landscape contractor
   b. landscape architect
   c. landscape gardener
   d. landscape designer

5 The process of removing selected branches of trees and shrubs is called
   a. balancing
   b. pinching
   c. pruning
   d. tying

6 Planting designs that generally have a more formal, traditional appearance are
   a. asymetrically balanced
   b. symetrically balanced
   c. not balanced
   d. informally balanced

7 In the landscape maintenance business, the most time and money is spent annually on
   a. interiorscaping
   b. pruning
   c. irrigation
   d. lawn care

8 Most professional pruning occurs in the
   a. spring
   b. summer
   c. fall
   d. winter

9 Woody landscape plants that lose all of their leaves at one time during the fall are
  a. deciduous
  b. evergreen
  c. perennials
  d. annuals

10 To avoid damaging other plants when installing a landscape design, it is important to first plant
  a. shrubs
  b. ground covers
  c. sod or grass seed
  d. trees

Discussion

1 Explain what is meant by landscaping.

2 What are the five basic landscaping design principles?

3 How is unity expressed?

4 What is a landscaping plan?

5 What is meant by landscape maintenance?

CHAPTER 9

Fruit and Nut Production

Student Objectives

When you have finished studying this chapter, you should be able to:

- Explain how consumers can buy fresh fruit all year long.
- List the major fruit and nut crops in the United States.
- Describe how pome fruits are produced.
- Describe how stone fruits are produced.
- Describe the citrus industry.
- Discuss nut production in the United States.

Key Terms

pome fruits	cultivars	citrus fruits
grafting	stone fruits	

FIGURE 9–1
Many types of fresh fruit can be bought all year long. *Courtesy of James Strawser, The University of Georgia.*

I f you go into any large grocery store you can find a wide variety of fruit. Fresh fruit of many types is readily available. Apples, peaches, cherries, oranges, strawberries, pears, bananas, and many other fruits can be bought at any time of the year (Figure 9–1). This has not always been the case. Fresh fruit spoils very easily. Until relatively recently, people could only eat fresh fruit during the time between when the fruit ripened and when it spoiled. Fruit was preserved, but preserved fruit was never the same as fresh. The fruit was dried, canned, or made into jelly and preserves. Even though a lot of fruit is eaten fresh, much fruit is still preserved by these methods (Figure 9–2).

Modern transportation and storage techniques allow us to move fresh fruits long distances. In the southern hemisphere, the seasons are opposite from ours. When it is winter in North America, it is summer in that part of South America that is south of the equator. When our fruit trees are dormant, theirs are in full production. Fruit is shipped from South America and we can have fresh fruit in the middle of the winter. Peaches, pears, plums, and

FIGURE 9–2
Much fruit is still preserved. *Courtesy of James Strawser, The University of Georgia.*

FIGURE 9–3
Apples do best in climates where the nights are cool during ripening. *Courtesy of James Strawser, The University of Georgia.*

grapes are shipped to the United States during the winter. Most of the fresh apples we enjoy have been stored from the U.S. crop.

Pome Fruits

Pome fruits are those that have fleshy insides with small seeds. Only two pome fruits of importance are grown in this country—apples and pears. Most of the apples are grown in Washington, New York, Michigan, California, and the higher altitudes of the Appalachians. Apples grow so well in these areas because the trees need a cool climate. Cool nights during the ripening process make the apples have a good flavor and color (Figure 9–3). In the South, the nights are warmer during the time when the fruit ripens and the quality of the fruit is not as good.

About half of the apples produced are marketed fresh. The rest are processed into juice, jellies, pies, and other products. The most popular varieties are Red Delicious and Yellow Delicious, which are sold as fresh fruit. Other varieties include Macintosh, Rome Beauty, and Jonathan. These varieties are usually used for processing or cooking.

Most of the pears in this country are grown in Washington, Oregon, and California (Figure 9–4). Pears do well in the semiarid regions of these states. Pears are more difficult to produce in the East because of disease. A disease called fire blight attacks pear trees in the more humid areas. This blight is difficult to control and makes the pears expensive to produce. Many of the same practices used in the growing and managing of apples are used in growing pears.

Apple Production

The first step in the production of apples is the production of young trees. Instead of growing the trees from seed like many other plants, a process called

FIGURE 9–4
Most pears are grown in Washington, Oregon, and California. *Courtesy of Washington State University.*

FIGURE 9–5
Pruning helps produce a more desirable tree. *Courtesy of James Strawser, The University of Georgia.*

grafting is used. Apples are difficult to propagate by any other method. For example, a hybrid apple tree might be difficult to grow from seed because the tree will be different from either of its parents. If a young sprout is combined with a portion of a desirable tree, the results can be a tree that grows and produces like the desirable tree (Figure 9–5). This method is also used to produce specialty trees such as dwarf trees and fruit trees that will grow several different varieties of fruit (Figure 9–6). Trees are also grafted onto rootstock that is stronger or more suited to the area.

Grafting involves cutting two parts of a tree. The lower part is called a rootstock and the upper part is called a scion (sometimes spelled "cion"). The rootstock may be larger than the scion. This gives the advantage of faster growth and earlier maturity. The two cuttings are carefully aligned and wrapped in string. The joint is then coated with wax and the plant is set in the soil. Eventually the two plant parts grow together and form a new tree. Through grafting, several varieties of apples

FIGURE 9–6
Through grafting, several varieties can be grown on the same tree. *Courtesy of USDA.*

can be grown on the same tree. Scions from different varieties are grafted on to different branches of a tree. Each grafted limb will grow the same type fruit as the tree from which the scion grew.

The young trees are set out in evenly spaced rows. Setting out in rows provides room for mowing and spraying machinery to maneuver. In arid areas, irrigation systems may be installed. The young trees are kept free of weeds and sprayed to prevent damage from insects and disease. Also, the young trees are pruned to shape them and to prevent the growth of weak branches.

It generally takes several years for the trees to begin bearing fruit. The length of time varies according to the variety, climate, soil conditions, management practices, and size of the trees. Because of the length of time involved before the trees produce, producers have a large investment in the trees (Figure 9–7).

Some trees, especially certain **cultivars**, are self-sterile. Cultivars are plants that are cultivated and grown by humans. These plants may have been bred up to the point where they are sterile unless pollinated from other plants. Red Delicious apple trees are a good example. By themselves, they are sterile and need another apple tree to cross-pollinate. Yellow Delicious trees produce pollen that is excellent in fertilizing Red Delicious trees.

Fruit trees are generally pollinated by insects (Figure 9–8). Remember from Chapter 4 that honeybees are particularly adept at pollinating.

Fruit growers use bees to pollinate the trees. Hives are set in the orchards every year during the spring. The bees work the blossoms and also produce honey. Honey from fruit blossoms is delicious and is in great demand.

Once the trees have bloomed and the fruit has set, the trees have to be sprayed regularly to prevent

FIGURE 9–7
An orchard represents a sizable investment before fruit is produced. *Courtesy of James Strawser, The University of Georgia.*

FIGURE 9–8
Most fruit trees have to be pollinated by bees or other insects. *Courtesy of James Strawser, The University of Georgia.*

insect and disease damage. Both disease and insect damage lowers the quality of the fruit. Modern insecticides are safe when used correctly. The USDA carefully monitors the pesticide residues on fruit.

When the apples are ripe, harvesting begins. Most apples are harvested by hand to make sure the fruit is not bruised. Bruises lower both the appearance and flavor of the fruit. Pickers are careful how they handle the fruit (Figure 9–9). The fruit is gathered in small containers that do not damage the fruit.

Apples that are to be stored as fresh fruit receive special handling at the processing plant. Once the fruits are washed and cleaned, they are coated with a thin coating of paraffin to prevent damage to the skin and to help seal the surface of the fruit. The apples are stored in a process called controlled atmosphere storage. The fruit is placed in dark chambers where the temperature is held at a constant temperature of between 32 and 55 degrees F. Oxygen and carbon dioxide levels are decreased

FIGURE 9–9
Pickers have to be careful not to damage the fruit. *Courtesy of James Strawser, The University of Georgia.*

and are replaced by nitrogen. The apples are maintained in a proper balance of oxygen, carbon dioxide, and nitrogen. These gases, combined with the correct cool temperature and the absence of light, allow apples to be kept fresh for many months (Figure 9–10).

Stone Fruits

Stone fruits have large seeds called pits or stones in the center. The most commonly grown stone fruits are peaches, cherries, plums, and apricots. All of these fruits are grown worldwide and in the United States.

Peach Production

Peaches are grown all across the country (Figure 9–11). The major states are California, South Carolina, Georgia, New Jersey, Pennsylvania, and Michigan. Peaches need a certain number of hours of cold weather to bring the trees through the process of dormancy. Most varieties need around 750 hours of weather below 45 degrees F in order to have the proper amount of blossoms and fruit.

FIGURE 9–10
Properly stored, apples can be kept fresh for a year or more. *Courtesy of Henry Waelti, Washington State University.*

There are two types of peaches. One type is known as freestone. These varieties of peaches have stones that easily separate from the flesh when the peach is cut in half. The other type, clingstone, does not separate from the flesh of the peach. Most of the peaches that are marketed fresh are freestone and most clingstone are used for processing.

Like most fruit trees, peaches are grafted onto rootstock. They are set out in rows and managed by pruning, fertilizing, and spraying. Peach trees are pruned differently than apples. Peach trees are pruned to have an open, inverted umbrella shape. This is to allow more sunlight into the branches of the tree (Figure 9–12).

FIGURE 9–11
Peaches are grown all across the country. Different varieties grow in different areas. *Courtesy of James Strawser, The University of Georgia.*

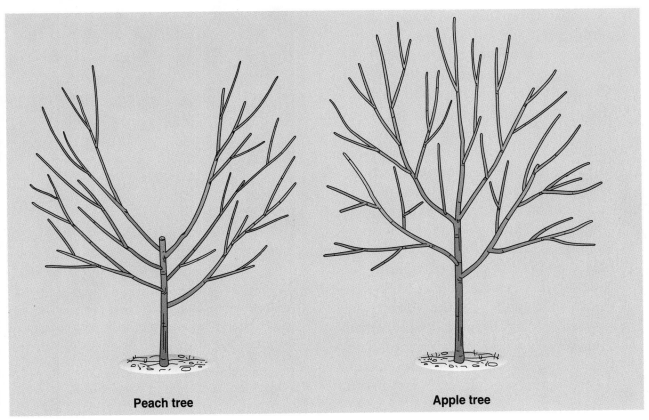

Peach tree   Apple tree

FIGURE 9–12 Peach trees are pruned to have an open center.

Most peaches are harvested by hand. The fruit is so soft that it bruises very easily. Fruit that is intended for the fresh market must be free of bruises and blemishes. Peaches that are intended for processing may be harvested by a mechanical picker. These machines have a flexible platform that goes around the tree. The machine grabs the tree and gently shakes it. Ripe fruit falls onto the platform and the peaches are gathered in. The fruit is rushed to market where it is washed, peeled,

Career Development Events

Everyone likes to receive recognition. The FFA Proficiency Awards and Supervised Agricultural Experience Program (SAEP) are two ways every student can participate and become winners. To begin, you must first become an FFA member and be enrolled in an agriculture class. Your teacher will help you design an SAEP that will put into practice your interests plus the skills and abilities learned in the classroom, shop, and laboratory. By becoming active in an SAEP, you will learn how to earn money that you may reinvest, spend, or save. You will also develop the skills required to produce, operate, and market your selected product. The Fruit and Vegetable Production Proficiency Award recognizes individual FFA members who develop and manage SAEPs in producing and marketing products such as field grown vegetables, orchard fruit,

The Fruit and Vegetable Production Proficiency Award recognizes individual FFA members who excel in producing and marketing fruits or vegetables. *Courtesy of Blane Marable, Morgan Co. High School, Madison, GA.*

and processed. The processing may include canning, pie production, or making into preserves.

S t o p

Citrus Fruit

Citrus fruits have been grown for thousands of years. They are thought to have originated in Southeast Asia. They were introduced to Europe in the 12th century and were brought to America by the European settlers. These fruits provide us with a

and vineyard crops. Recordkeeping is a very important part of a Supervised Agricultural Experience Program (SAEP). The student must maintain a list of fruit and/or vegetable production skills that are learned in the classroom and/or used during the SAEP. Development of equipment and material inventories plus the recording of income, expenses, and accomplishments must be accurately kept. The student must make managerial decisions involving the production and marketing of the fruit and/or vegetables and also provide the labor in planting, maintaining, and harvesting. An award-winning proficiency is the result of a long-term SAEP, good record-keeping, and a student's involvement with the local FFA and community.

Kerry won the Fruit and Vegetable proficiency award in his FFA chapter. He had grown a small vegetable garden his first year in agriculture and won the chapter garden contest. In his second year he rented one acre of land and raised tomatoes and squash and sold them at the local farmer's market. Kerry added the production of cool season crops (lettuce and cabbage) his third year of agriculture. After the fourth year of his SAEP, Kerry had made enough income to purchase two acres, where he plans to continue his vegetable business. Kerry was on his high school football team and chaired the FFA Food for America program for two years. While attending the state FFA convention, Kerry placed second in the tractor driving contest. Kerry's experiences with the FFA and his SAEP have prepared him to be a successful leader, businessperson, and citizen.

FIGURE 9–13
Citrus is grown in the extreme southern areas of the United States. *Courtesy of Vick Christ.*

good tasting, nutritious addition to our diets. Citrus is an important source of vitamins and minerals, particularly vitamin C. During the 1750s, a British doctor named James Lind noticed that sailors who spent long tours at sea frequently came down with a disease called scurvy. The sailors were fed on a diet of dried beans and dried meat. Dr. Lind discovered that by giving the sailors citrus fruit, scurvy could be avoided. Later, it became known that the vitamin C in the fruit prevented the disease.

The citrus grown in the United States includes oranges, grapefruit, tangerines, tangelos, lemons, and limes. Most of the citrus is grown in Florida, California, Texas, and Arizona. These areas are about the only places in the United States with a climate warm enough to produce the fruit (Figure 9–13).

Although much of the crop is sold as fresh fruit, there is a large industry that processes citrus. The largest processed products are orange, grapefruit, and lemon juices. The fruit is crushed to remove the juice. Much of the water is removed from the juice, and it is packaged and frozen. The consumer can then thaw the concentrate, add the proper amount of water, and enjoy orange juice (Figure 9–14).

FIGURE 9–14
Citrus juice is concentrated, then frozen. *Courtesy of James Strawser, The University of Georgia.*

One interesting by-product of the orange and grapefruit juice industry is the pulp. Pulp is the fibrous material left over when the juice is pressed out of the fruit. This material makes a valuable cattle feed. When mixed with grain, citrus pulp is used to fatten cattle for market. A good use has been found for a material that once went to waste.

Nut Production

Nuts are actually seed kernels that are housed in a woody structure. They have been used as food for as long as people have been on the earth. Most nuts are high in proteins and fats. Nuts may contain as much as 600 to 700 calories per 100 grams compared to 250 calories per 100 grams for beef with moderate fat content.

Each year producers in the United States harvest over 430,000 tons of nuts. The nuts are grown in the warmer regions of the country. While nut trees will grow in all states, they are generally sensitive to the cold and grow better in the South. Four types of nuts are of real commercial importance in the United States: almonds, pecans, walnuts, and filberts.

More almonds are produced in the United States than any other nut. Almost all of the almonds grown in this country are produced in California (Figure 9–15). The almond is native to the hot, arid regions of West Asia and California matches the climate. The almond tree resembles the peach tree. In fact, peach trees are often used as rootstock for grafting almond trees.

Although there are over 100 varieties of almonds grown, there are two major types of almonds. The sweet almond is edible and is grown for food. The bitter almond is grown for the oil in the kernel. The sweet, edible almond is eaten raw, roasted, or in candies and pastries. The oil from the bitter

FIGURE 9–15
Almost all almonds grown in the United States are grown in California. This machine is shaking the nuts from the tree. *Courtesy of Jack Kelly Clark, University of California, Statewide IPM Project.*

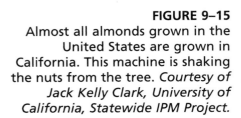

almond is used to produce prussic acid that is valuable as a fumigant. The oil is also used in some medicines.

The major type of walnut is the English walnut that is grown in California and Oregon (Figure 9–16). The black walnut is also grown, but it has smaller kernels that are more difficult to remove from the hull. Most English walnuts are grafted onto black walnut rootstock because the black walnut trees have stronger root systems.

FIGURE 9–16
English walnuts are grown in California and Oregon. *Courtesy of Jack Kelly Clark, University of California, Statewide IPM Project.*

Most walnuts are sold in the shell, although some are sold as kernels. These shelled nuts are used for baking and in salads. Another use for the walnut trees is the valuable wood. Walnut has always been a highly prized furniture wood.

Pecans are classified as genus *Carya*, which includes the hickories. Even though this nut tree is classified as a hickory, the nuts differ from other hickory nuts. Pecans have larger kernels and thinner shells. In fact, some pecan varieties are known as "paper shell pecans" because of the very thin shells.

Pecans are grown in the South. The leading states are Georgia, Alabama, Texas, Louisiana, New Mexico, and Oklahoma. The trees grow quite large (75 to 90 feet) and mature slowly (Figure 9–17). Most need eight to ten years from the time they are set out until they begin to produce. The trees live for many years and produce a long time.

Pecans are sold both shelled and unshelled. The nuts are used in pastries, pies, and other sweets.

FIGURE 9–17
Pecan trees grow quite large and mature slowly. *Courtesy of James Strawser, The University of Georgia.*

FIGURE 9–18
Nuts are removed from the tree by a machine that gently shakes the tree. *Courtesy of James Strawser, The University of Georgia.*

They are relatively high in fat content and contain a considerable amount of protein.

Filberts are also known as hazelnuts. Most are grown in Oregon and Washington. They are eaten raw or roasted and are becoming increasingly popular. One of the newer marketing techniques is to make consumers aware of the fact that filberts are hazelnuts. Consumers generally do not recognize the name filberts. The nuts grow on either a small tree or shrub depending on the management practices used. The trees bloom in the winter, and the nuts mature in the fall.

Nuts are harvested by using a tree shaker. The machine grabs the tree and gently shakes it until the nuts fall off (Figure 9–18). Another machine sweeps the nuts into rows called windrows. A large vacuum picks the nuts off the ground (Figure 9–19). This means that the ground beneath the trees needs to be kept clean of weeds and foreign objects such as rocks. At the processing plant, the nuts are shaken to remove rocks and limbs. They are then

FIGURE 9–19
The nuts are picked up from the ground by a nut sweeper. *Courtesy of James Strawser, The University of Georgia.*

placed in large bins, where the excess moisture is removed.

Summary

The fruit and nut industry in the United States is very large. This industry provides consumers with food that is delicious and nutritious. Our lives are made better through the availability of fruits and nuts all year round. In addition, they greatly add to our health.

9 CHAPTER REVIEW

Student Learning Activities

1. Visit a large grocery store and make a list of all the different types of fresh fruit that are available. Determine what country produced the fruits. Compare your findings with others in the class.

2. With your teacher's assistance, slice up several different varieties of apples and ask people in the class to taste the samples. Keep a list of the ones favored by the most people. Ask each person to tell why he or she preferred a particular variety.

3. Survey your area and determine what types of fruit and nuts are grown in your area. Report on the climatic conditions and other factors that make your area good for producing these crops.

4. Go to a large grocery store and find fruit you have never tried before. Purchase a sample and try it. If it is not a well-known fruit, try to explain why it is not more popular in your area.

True/False

1 ___ Fresh fruit spoils very easily.

2 ___ Most of the fresh apples we enjoy come from South America.

3 ___ Apple trees do best in cool climates.

4 ___ Fruit trees are generally pollinated by insects.

5 ___ Peaches, cherries, plums, and apricots are examples of stone fruits.

6 ___ Citrus pulp makes a good cattle feed.

7 ___ Nuts are grown in the cooler regions of the country.

8 ___ More pecans are produced in the United States than any other nut.

9 ___ Hazelnuts are also known as filberts.

10 ___ Nuts are harvested by using a tree shaker.

Multiple Choice

1 The only two pome fruits of importance that are grown in this country are apples and
   a. oranges
   b. pears
   c. cherries

2 The most popular variety of apple is
   a. Granny Smith
   b. Jonathan
   c. Red Delicious

3 Fruit growers make use of bees to pollinate the
   a. flowers
   b. seeds
   c. roots

4 The two types of peaches are freestone and
   a. filberts
   b. sandstone
   c. clingstone

⑤ Most of the peaches that are marketed fresh are
    a. freestone
    b. clingstone
    c. sandstone

⑥ Citrus fruits are thought to have originated in
    a. North America
    b. South Africa
    c. Southeast Asia

⑦ Most nuts are high in fats and
    a. filberts
    b. proteins
    c. seeds

⑧ Almost all of the almonds grown in the United States are produced in
    a. Washington
    b. Georgia
    c. California

⑨ The almond tree resembles the
    a. peach tree
    b. apple tree
    c. walnut tree

⑩ Pecans are grown in the
    a. West
    b. North
    c. South

Discussion

① In the past, people could only eat fresh fruit during a very brief time of the year. Why?

② What are some ways fruit is used?

③ Why are cool nights beneficial to apples?

④ What is fire blight?

⑤ What are the two types of peaches?

⑥ Why are peach trees pruned into an inverted umbrella shape?

7 How was the importance of vitamin C in citrus fruits discovered?

8 In the United States, where is most citrus fruit grown?

9 What three citrus fruits are processed most often?

10 What are the two types of almonds?

CHAPTER

Row Crops

Student Objectives

When you have finished studying this chapter, you should be able to:

- List the major row crops grown in the United States.
- List the types of grain that are produced in the United States.
- Describe the production and uses of grain crops.
- List the types of oil seed crops that are produced in the United States.
- Describe the production and uses of oil seed crops.
- List the types of fiber crops that are produced in the United States.
- Describe the production and uses of fiber crops.

Key Terms

row crops	flour	oil crops
grain	cereal	fiber crops

3 Factors that determine what kind of row crops are grown in a state ←

A large portion of our gigantic agricultural industry is row cropping. **Row crops** are those that are planted and cultivated in rows. Usually it refers to annual plants that have to be replanted each year. Every state in the country grows row crops. The type varies with the climate, soil type, and available markets. Some crops, such as corn, are grown in all 50 states.

Hundreds of different types of crops are grown in this country. They may be classified as grains, oil crops, and fiber crops. It would be impossible to describe all of the different types of crops grown in this country. However, a few major crops comprise most of the row crops in this country. These major crops will be described in this chapter.

Grain Crops

Probably the very first crop that people planted was **grain**. Grain is the seed from plants that are classified as grasses. Humans discovered that not only were the grass seeds good to eat, but they could be planted and more grass that produced more seed would grow. The grasses that produced the most and best seed were selected for cultivation (Figure 10–1). As more was grown year after year, plants were developed that produced more seed that could be used for food.

Wheat

Wheat is perhaps the oldest of the major grain crops. Ancient religious documents mention the growing of wheat. Most civilizations have grown wheat to be used in the making of bread. Bread is a staple food that can be preserved and provides nutrition.

Most of the wheat grown is used for human food. It has been the basis of bread for thousands of

FIGURE 10–1
Grasses that produce the most and best seed are cultivated as grain. *Courtesy of James Strawser, The University of Georgia.*

years. The seed kernels are ground into **flour**, and flour is used in baking. Wheat flour contains a high percentage of a substance called gluten, which causes the flour particles to stick together when moisture is added (Figure 10–2). This sticky dough traps gases from fermentation that cause the bread to rise. When compared to other grains, wheat flour is relatively high in protein and makes a good-tasting food.

FIGURE 10–2
A substance in wheat flour called gluten makes it good for baking bread. *Courtesy of James Strawser, The University of Georgia.*

Wheat is grown in many countries of the world. It can be grown in a variety of climates including cool climates. The United States produces about 12 percent of the world's wheat.

Wheat is grown all across the country, but most of the production is centered in the Midwest and Great Plains. Several types of wheat are grown in this country. These include hard red winter, soft red winter, hard red spring, white, and durum wheat.

Hard red winter wheat is grown in Kansas and nearby states. More hard red winter wheat is grown than any other type. It contains a relatively high percentage of gluten that causes the flour to be among the best for baking breads. It is second only to hard red spring wheat for baking breads.

Soft red winter wheat is grown in areas of the Midwest where there is abundant rainfall. The moisture causes the wheat grains to be softer, more starchy, and lower in protein than the hard red winter wheat. Consequently, gluten content is lower and the baking quality of the flour is not as good.

Hard red spring wheat has the highest amount of gluten of the commonly grown wheats. The highest quality breads are made from hard red spring wheat (Figure 10–3). This wheat is grown in the upper Great Plains area of North and South Dakota, Montana, and Minnesota. The weather gets too cold in these areas to grow winter wheat, so much of the wheat production is spring wheat.

White wheat is grown in the states of Washington and Oregon with some being grown in Michigan and New York. White wheat is grown both as winter and spring wheat. The flour from white wheat does not make very good bread and is used mostly for baking crackers, pastries, and breakfast **cereal** foods.

Durum wheat is grown through the upper Great Plains. It is more drought- and disease-resistant than the hard red winter wheat. Durum wheat

FIGURE 10–3
Hard red spring wheat makes the best flour for bread. *Courtesy of Getty Images.*

contains a high percent of gluten but it is considerably lower quality than the hard red winter and spring wheats. Durum is used for making pastas such as macaroni and spaghetti.

Winter wheat is planted in the fall and harvested in mid- to late summer. Spring wheat is planted in the early spring and harvested in the late summer and early fall. At planting, the seeds are placed in the ground using an implement called a grain drill (Figure 10–4). The drill places seeds in close rows (6 to 12 inches apart) at the proper depth in the soil. Winter wheat must go through a period of prolonged cold before it can form grain heads. This process if called vernalization.

Wheat harvesting is done with a combine (Figure 10–5). This machine cuts the wheat stalks off and threshes the grains of wheat out of the seed heads. A large fan blower separates the hulls and chaff from the grain. The grain is then conveyed to a bin that when full is emptied into a truck. The truck takes the grain to an elevator where it is sampled, graded, and stored. The producer is paid on the

FIGURE 10–4
Wheat is planted using an implement called a grain drill. *Courtesy of Dewey Lee, The University of Georgia.*

FIGURE 10–5
Grain is harvested using a combine.
Courtesy of James Strawser, The University of Georgia.

basis of the grade of the wheat. The grade is based on moisture content and the amount of weed seed and other foreign matter in the grain.

Corn

More corn is produced in the United States than any other crop. The United States produces almost half of all the corn produced in the world. It is one of the few major crops that is native to this country. Before the European settlers arrived in the New World, corn was not grown in other parts of the world. However, it was widely cultivated by Native Americans in both North and South America.

Corn is grown in every state in the country, but the majority of the production is in the Midwest (Figure 10–6). In this region, the soil is deep and rich and the rainfall is adequate. This combination allows for huge yields of corn. These yields have dramatically increased over the past few decades. Research into methods of production has greatly increased production. However, the greatest impact has come from hybrid varieties. A hybrid is a cross between two different varieties. The resulting plant outproduces either of the parents. In 1930, when

FIGURE 10–6
Corn is grown in every state in the United States. *Courtesy of James Strawser, The University of Georgia.*

very little hybrid corn was grown, the average production was around 25 bushels per acre. Today, when almost all corn is hybrid varieties, the average yield is almost 150 bushels per acre. This average is for all of the country. A well-managed field in the Midwest may average well over 200 bushels per acre (Figure 10–7).

Corn is used both for human food and animal feed. Almost all of the corn produced is either white or yellow. Of the two types, yellow corn is by far the most widely grown. White corn is grown for corn meal that is used in baking and bread making. It is also used in making breakfast cereals and corn chips because it toasts better than yellow corn. Yellow corn contains a substance called carotenoid pigments that animals can convert to vitamin A. Both yellow and white corn are used for animal feed because they are so high in carbohydrates (Figure 10–8). Cattle and hogs are fattened on corn because it is so high in this nutrient. Because corn is relatively low in protein, a protein supplement is usually added to livestock feed.

Corn has other uses. Cooking oil is pressed from the grain. It has no cholesterol and is of very high quality. Corn starch is produced from the starchy

FIGURE 10–7
A well-managed field of corn in the Midwest may average as much as 200 bushels of corn per acre.
Courtesy of John Deere & Company.

content of the grain. It is used in thickening agents in puddings and pies, as a drying agent in body powder, and for a broad array of other uses.

A major new use of corn is the production of ethanol alcohol. Alcohol made from corn and other grains can be made to run car, truck, tractor, and

Career Development Events

FFA members have a great opportunity to compete for awards in the area of row crops, including grain, oil, and fiber products. These awards encourage the use of new technology and management practices in the production and marketing of agriculture crops. By taking agriculture classes, the student can learn and practice the knowledge and skills required to become an effective producer of agriculture products. Producing crops is a great Supervised Agricultural Experience Program (SAEP) for any FFA member. The Agriculture teacher, student, and guardian should agree on the scope and focus of any SAEP. Students will be responsible for making managerial decisions about the ground preparation, planting, maintenance, harvesting, and marketing of the crop. The student must keep a list of crop production skills used in the classroom and in the field. Income, expenses, and yields of the crops produced and harvested also must be accurately recorded. The building of inventory such as land, seed, fertilizer, machinery, and harvested crops over several years is essential to an award-winning proficiency.

Jill is an active member of her FFA. In her senior year she won the state agriculture management contest and a scholarship to her state's agriculture college. Jill began her SAEP in the ninth grade. She lived on a 100-acre family farm, and part of the signed agreement between herself, her father, and agriculture teacher was to lease two acres of land to Jill. The first two years Jill tilled, sowed, managed, and harvested cotton on the two acres. Jill's agribusiness class taught her record keeping skills and investment opportunities that helped her reinvest her income wisely. Jill chose to sell her cotton with a local co-op. The last two years Jill diversified her production by crop rotations. After paying all the expenses of her operations, Jill's profit margin was small, but the experience was priceless. Jill's application for the crop proficiency award won chapter, district, and state recognition. Jill is one

other engines as efficiently as gasoline with much less pollution of the environment (Figure 10–9). The technology for this process is hundreds of years old and is really a simple procedure. Alcohol produced from corn holds promise for replacing much of the petroleum-based fuels used today. At our high rate

of many who have found the FFA to be the foundation of her future.

Frank is a high school student who lives in the city. This year he is applying for the chapter placement in agricultural production proficiency. Frank has been employed on a forage production farm for the last three years. Frank's FFA membership started in the seventh grade. He continued to be an active member and competed on the forestry and land judging teams. His sophomore year Frank came to his teacher looking for an after-school job. Frank's desire to be outdoors and his accomplishments in agriculture mechanics class led his teacher to place him on a local forage production farm just outside the city. Frank has kept accurate records of his hours of work, job skills, income, and expenses related to his job. Through the years, Frank was promoted from field hand to large equipment operator. His experiences and skills have prepared Frank for a career in agriculture, and he is an excellent candidate for the proficiency placement award in agricultural production.

Crop production has been a part of FFA activities for years. Students can still receive awards in this area. *Courtesy of Blane Marable, Morgan Co. High School, Madison, GA.*

FIGURE 10–8
White corn is used more for human food. Yellow corn is used more for animal feed. *Courtesy of James Strawser, The University of Georgia.*

of usage, petroleum oil will someday run out. A mixture of ethanol and gasoline called gasohol now replaces over 40 million barrels of crude oil each year.

Stop!

Oil Crops

A major part of the row crops in this country is produced for their oil content. The two most widely produced **oil crops** are soybeans and peanuts. The oil from the seeds of these plants is used in a

FIGURE 10–9
Some engines make use of ethanol that is made from corn. *Courtesy of Illinois Research Agricultural Experiment Station.*

FIGURE 10–10
Oil from the seed of plants is used for cooking oil. *Courtesy of James Strawser, The University of Georgia.*

variety of ways ranging from cooking oils to paints to a replacement for diesel fuel (Figure 10–10).

Soybeans

Soybeans probably originated in China, where they have been grown for thousands of years. They were imported into this country in the late 1800s. Through the development of varieties suited to the climate of various parts of the United States, soybeans have become a major crop in this country. In fact, the United States produces over half of the world's supply of soybeans (Figure 10–11). It is interesting that the United States now exports soybeans to China, the country in which the crop originated. This crop is grown extensively throughout the South and Midwest.

One reason that soybeans are popular is the wide variety of uses. One of the primary uses is for oil. The oil is pressed out and used for cooking and making margarine, shortening, cheese spreads, and coffee whiteners. New uses for the oil, such as a substitute for diesel fuel and as the basis for a high quality ink, have been discovered (Figure 10–12). The whole bean can be used to make meat substitutes and extenders.

FIGURE 10–11
The United States produces over half the world's supply of soybeans. *Courtesy of James Strawser, The University of Georgia.*

FIGURE 10–12
Soybeans are used for a variety of purposes ranging from animal feed to printer's ink. *Courtesy of James Strawser, The University of Georgia.*

The material left when the oil is pressed out is used for animal feed. The pressed cake is ground into a meal that contains an average of 45 percent protein. This high-quality feed is mixed with corn and other grains and fed to pigs and chickens.

Peanuts

The peanut is native to South America and was transported to Africa by Europeans. From there, it was brought to North America by Africans who were brought in as slaves. This nutritious food gained popularity in the latter part of the 1800s when the boll weevil destroyed much of the cotton crop of the South. Scientists such as George Washington Carver invented new uses for the peanut. Products such as peanut butter became popular and a new industry grew up in the South.

Today, this crop is grown in many parts of the South. The leading states are Georgia, Alabama, Texas, and North Carolina. The nuts grow on "pegs" that extend from the stem into the ground (Figure 10–13). They are planted in rows about 3 feet apart. At time of harvest, the vines may completely cover the ground. The plants require a long

FIGURE 10–13
Peanuts grow from "pegs" that extend from the nodes of the stem into the ground. The pods grow underground. *Courtesy of James Strawser, The University of Georgia.*

growing season with warm weather. In addition, because the nuts grow in the ground, they must be grown in light sandy soil. Peanuts grown in soils with heavy clays are difficult to harvest.

At harvest, a large plow turns the roots toward the sky (Figure 10–14). The vines are left in the field for several days until the nuts begin to dry. A large machine picks up the vines and removes the nuts. The nuts are allowed to dry and are stored or processed.

Peanuts are sold in the shell raw, shelled raw, roasted in the shell, shelled and roasted, or boiled. Products such as peanut brittle, a variety of candies, and pastries are made from peanuts. Of course, many of us enjoy eating peanut butter, which is made from ground up roasted peanuts.

One of the most valuable products from peanuts is peanut oil. This is one of the highest quality cooking oils. Like soybeans, peanuts are pressed by machines until the oil is removed. The cake that is left is used as a high-quality livestock feed. Not only is it high in protein content (around 50 percent), but also the protein is very high quality. In addition, the dried plants are often fed to cattle as hay.

FIGURE 10–14
A large plow turns the peanuts toward the sky. *Courtesy of James Strawser, The University of Georgia.*

Fiber Crops

Much of the cloth produced today (such as rayon, nylon, and polyester) is manufactured from petroleum. In a process that was developed in the first half of the twentieth century, petroleum is processed into long fibers that are used to make cloth. This process gives us cloth that is relatively inexpensive and very durable. For these reasons, **fiber crops** declined in popularity for several years. However, today, natural fibers are rebounding as a clothing material. Human-made fibers may be more wrinkle resistant and durable but they cannot match fiber produced by agriculture for comfort (Figure 10–15).

FIGURE 10–15
Human-made fibers cannot match cotton clothing for comfort. *Courtesy of James Strawser, The University of Georgia.*

Also, a lot of cloth produced is a combination (called a blend) of artificial and natural fibers that gives the flexibility and durability of synthetics and the comfort of natural fibers. The use of natural fibers makes us less dependent on imported petroleum.

Two main kinds of natural plant fibers produced for cloth are grown in the United States. These are cotton and flax, and both of these are also used as oil crops.

Cotton

Cotton was probably first used in the Nile Valley in ancient Egypt. The climate and the fertile soil were ideal for the growth of the plant. In North America, humans have used cotton for at least 7,000 years, and its cultivation goes back hundreds of years before the European settlers arrived. Much of the history of the United States developed around the growing of cotton.

Cotton is a shrublike perennial plant (although it is cultivated as an annual plant) that requires a long growing season and warm temperatures

FIGURE 10–16
Cotton requires a long growing season. *Courtesy of James Strawser, The University of Georgia.*

(Figure 10–16). This means that the crop must be grown in the southern parts of this country. The leading states are California, Texas, Arizona, and the states of the lower Southeast.

About mid-season, the cotton flower blooms from a bud that appears on the stem. As it opens, the flower is a pale yellow that gradually turns pink and then red (Figure 10–17). After the flower is fertilized, the seed pod, known as a boll, forms. This structure is shaped like a small football and contains the fiber and the seed. The seed grows within the boll and the fibers form on the seeds.

Cotton is picked by large machines that use rotating fingerlike projections on opposite rotating drums to remove the lint from the bolls (Figure 10–18). This ingenious invention replaced long, strenuous hours of manual labor that was once required to handpick the cotton. The lint is removed from the picker and is either stored in the field in long loaf-shaped modules that hold 10 to 12 bales of cotton or it is placed in trailers and taken directly to the gin. A bale weighs around 1,200 pounds before the seeds are removed and around 500 pounds after the seeds are removed.

FIGURE 10–17
Cotton blossoms are first yellow, then turn pink. *Courtesy of James Strawser, The University of Georgia.*

FIGURE 10–18
Cotton is picked by large machines.
*Courtesy of James Strawser, The
University of Georgia.*

At the gin, the seeds are removed and the cotton is cleaned of trash and foreign material. Once the seeds are separated and the cotton is cleaned, the lint is pressed into 500-pound bales, graded, and stored or sent to the mill. The seeds are a secondary industry. They are sent to a processor where they are pressed to remove the oil. The oil is used as a cooking oil or in a variety of products from margarine to salad dressing. The cake that is left over after the oil is pressed out is ground into a meal that is a valuable source of protein. Cottonseed meal is fed to cattle and other ruminants. It cannot be utilized by pigs and certain other simple-stomach animals because of a toxic substance in the seeds called gossypol.

Flax

The use of linen dates back to the Stone Age when humans still wore skins for clothing but used linen for making fish nets. Over the years, linen has been used for clothing in most countries of the world. Much of its popularity as a fabric for clothes was lost with the widespread availability of cotton that began in the early 1800s. Today it is still popular as a cloth for making tablecloths, napkins, and in recent years has regained some popularity as a clothing material.

Linen comes from fibers produced in a plant called flax that grows in climates with plenty of rain and no high temperatures (Figure 10–19). Most of today's flax is produced in Europe and New Zealand. However, substantial amounts are produced in North and South Dakota and Minnesota.

Plant stems contain two structures, the xylem and the phloem, that transport water and nutrients through the plant. Linen comes from the fibers (called bast fibers) that make up the phloem of the plant.

The plants grow to a height of about 3 feet and are harvested. The outer layer of woody material must be dissolved before the fibers can be removed. This is done by soaking the stems in warm water where bacterial action decays the material or in a more modern method where the covering is dissolved using chemicals. This process is called retting.

After the outer layer is removed, the phloem fibers are passed through fluted rollers that break up the woody substances connected to the fibers. In a process called scutching, the usable fibers are separated out. Then the fibers are combed out in much the same process used in carding cotton. This lays out the fibers parallel to each other so they can be spun into yarn.

The seeds of the flax plant are pressed for their oil. This valuable oil, called linseed, is used in the manufacture of paints and varnishes. As with the other oil seeds, the cake left from oil removal is used for livestock feed.

FIGURE 10–19
A high-quality cloth called linen is made from the flax plant. *Courtesy of the Flax Council of Canada.*

Summary

The production of row crops in the United States is a gigantic business. No other country in the world can compare to the United States in terms of the amount or variety of crops that are produced. Much of the wealth of the nation is a result of our crop production. In the future, perhaps, new crops and new uses for old crops will be developed.

 # CHAPTER REVIEW

Student Learning Activities

1. Examine the labels on foods in your home. List all the foods that contain grain or grain products. Remember that gluten and starch come from grain. Compare your list with others in the class.

2. Visit a grocery store and list all of the products that are made from wheat. Try to determine which type of wheat would be best suited for each product.

3. List all of the products in your home that were made from oil from seeds. Try to think of other ways oil from plants might be used someday. Share your ideas with the class.

4. Determine the top row crops grown in your state. List reasons why these particular crops are the ones most widely grown.

True/False

1. ___ Probably the very first crops that people planted were grains.

2. ___ Most of the wheat grown is used for human food.

3. ___ Most wheat is produced in the Northeast.

4. ___ More corn is produced in the United States than any other crop.

5. ___ The majority of corn production is in the Midwest.

6. ___ Much of the cloth produced today is manufactured from petroleum.

7. ___ Human-made fibers are more comfortable than fibers produced by agriculture.

8. ___ Cotton was probably first used in the Nile Valley in ancient Egypt.

9. ___ One of the most valuable products from peanuts is peanut oil.

10. ___ China produces over half of the world's supply of soybeans.

Multiple Choice

1. A large portion of our agriculture industry is
   a. row cropping
   b. corn production
   c. fruit production

2. The grasses that produce the most and best seeds are selected for
   a. grazing
   b. cultivation
   c. flax

3. The oldest of the major grain groups is perhaps
   a. wheat
   b. rice
   c. corn

4. Wheat flour is relatively high in
   a. fat
   b. sugar
   c. protein

5. Gasohol is a mixture of gasoline and
   a. ethanol
   b. diesel fuel
   c. propane

6. The two most widely produced oil crops are soybeans and
   a. corn
   b. olives
   c. peanuts

7 A shrublike plant that requires a long growing season and warm temperatures is
   a. flax
   b. cotton
   c. linen

8 Linen comes from fibers produced in a plant called
   a. gluten
   b. cotton
   c. flax

9 Plant stems contain these two structures:
   a. xylem and phloem
   b. xylem and gluten
   c. phloem and gluten

10 The seed of the flax plant is pressed for
   a. gluten
   b. linen
   c. oil

Discussion

1 What factors determine what row crop is grown in each state?

2 What are the three classifications of crops?

3 What are the types of wheat?

4 Why is hard red winter wheat so popular?

5 Durum wheat is used for what purpose?

6 Other than color, what is the difference between white and yellow corn?

7 What are some uses for corn?

8 What functions are served by the xylem and phloem?

9 Where is the majority of corn grown? Why?

10 What uses are made of soybean oil?

Forest Science

Student Objectives

When you have finished studying this chapter, you should be able to:

- Discuss the extent of the forest industry.
- Explain why forestry can be considered agriculture.
- Explain the concept of natural succession.
- Discuss the forest as an ecosystem.
- Discuss the importance of the forests to humans.
- Discuss the production steps in growing trees.

Key Terms

forestry	softwood	wilderness areas
wildlife habitat	succession	tree farms
hardwood	canopy	

FIGURE 11–1
Like any crop, trees are planted, managed, and harvested. *Courtesy of Georgia Pacific.*

One of the largest industries in agriculture is **forestry**. At one time forests were considered natural resources. Giant forests once stretched across the country and people made use of the trees for building homes. When it became apparent that these forests would not last forever, the cultivating of trees began. Trees are now grown very much the same as crops are grown. They are planted, managed, and harvested (Figure 11–1). Once harvest is complete new trees are planted and a new crop is grown.

The forestry industry stretches from coast to coast and encompasses almost 500 million acres of trees. From these trees come thousands of different types of products that are manufactured using wood fiber. Over a million and a half people are employed in the management, harvesting, processing, and marketing of these products.

Some people think that the harvesting of timber is a horrible practice that destroys nature. This is an erroneous concept. Because of the conservation efforts of the forestry industry and the U.S. Forestry Service, today there are more trees in this country than there were 100 years ago. Each year close to two billion trees are planted in the United States. This represents more than six new trees a year for every American. Through the replanting of the forests, new plant growth flourishes. Each year about 27 percent more trees are planted than are harvested (Figure 11–2). Areas that once were used for row cropping are now used for growing trees. Hilly land not suited for farming with modern machinery is now planted in pine trees and is once again productive land.

Wildlife greatly benefits from the forest industry. As new trees are planted, lush growths of plants grow with the seedling trees. These plants provide food and habitat for wildlife. Several species of animals have been reintroduced where they had

FIGURE 11–2 Each year, about 27 percent more trees are planted than are harvested. *Courtesy of James Strawser, The University of Georgia.*

FIGURE 11–3
Today, there are more of some types of wildlife in the forests than ever before. *Courtesy of Vann Cleveland,* Progressive Farmer.

ceased to exist (Figure 11–3). For example, the white-tailed deer has made a tremendous comeback in recent years. Some sources say that there are now more deer in this country than before the first European settlers arrived. A few decades ago, the wild turkey was on the verge of extinction because of the loss of its habitat and overhunting. Today, there are over four million wild turkeys living and thriving in our forests.

Most of the trees that are harvested in the United States were planted or have grown up from forests that were cut. Only in the Pacific Northwest are there large tracts of old-growth timber. This timber is also call virgin timber and has never been cut. Even today, timber companies are cutting old-growth timber in the Pacific Northwest. No finer construction timber exists. Homes all over the world are made from lumber cut from the trees in the Pacific Northwest. Many people object to cutting the huge old trees (Figure 11–4). A constant struggle seems to be ongoing between preservationists and the logging industry. The preservationists insist that as much of the old-growth timber

FIGURE 11–4
Some object to the cutting of large, old-growth trees. *Courtesy of Jim Rathert, Missouri Conservation Department.*

FIGURE 11–5
Logging operations were stopped in efforts to save the spotted owl. *Courtesy of Northwest Forestry Association and Evergreen Foundation.*

as possible must be preserved. The logging industry points out that there are over four million acres of old-growth timber preserved in this country that, by law, can never be cut. They contend that the country's housing industry needs the material from the large trees that are in the areas where harvesting is permitted. They further argue that if logging is stopped many jobs will be lost as a result.

Another point of controversy is over **wildlife habitat**. A good example is the spotted owl that lives in the Pacific Northwest. Some biologists contended that the owl needed large tracts of old growth timber in which to breed and raise their young. Logging operations were halted and many people lost their jobs in the effort to preserve the owl (Figure 11–5). The issue was debated over whether or not preserving the owl was more important than providing jobs. Later evidence showed that the owls could reproduce in second-growth forests.

The Natural Forest

At one time, trees cut for lumber came from the natural forest. Different types of forest grew in different regions of the country. Basically, there are two classifications of timber: **hardwood** and **softwood**. A hardwood is a tree that is a broad leaf deciduous (sheds its leaves in the winter) tree (Figure 11–6). A softwood is a cone-bearing, evergreen tree called a conifer (Figure 11–7). Actually, the classification has little to do with the hardness or softness of the woods. Some hardwoods, such as yellow poplar, are softer than softwoods like southern pine.

The upper regions of the Northeast at one time were covered in vast hardwood forests. Throughout the Midwest and the upper regions of the South, hardwoods also grew. Along the coastal plains of the South, the predominate trees were pines. The Rocky Mountain region was covered with conifers such as

FIGURE 11–6 A hardwood is a deciduous tree. *Courtesy of James Strawser, The University of Georgia.*

FIGURE 11–7 A softwood is an evergreen, cone-bearing tree. *Courtesy of James Strawser, The University of Georgia.*

lodge pole and ponderosa pine. On the West Coast, gigantic softwoods such as Douglas fir, spruce, and redwood trees covered the slopes.

Forest Succession

The reason certain types or species of trees dominate a region is because of a natural process called **succession**. When a forest first begins from bare ground as a result of fire, clear-cutting, or other causes, a lot of different plants germinate. These plants all compete for water, nutrients, and sunlight. The trees grow taller than other plants so eventually trees shade the other plants. Plants that do not receive enough sunlight generally stop growing and die. Trees that grow well in a particular climate are usually the ones that grow faster than

Career Development Events

The Forestry Field Day Career Development Event is a team competition. Team members train for skills used in today's forestry industry (see photo). Tree planting, forest management, tree disorders, timber stand improvement, and estimating pulpwood and saw timber volumes are a few of the skills. Many forestry management practices are included in the FFA Forestry event. Contestants must understand and answer forestry management questions concerning stand classification, merchantable trees, basal area, and cutting practices. Team members must practice and be ready to give estimations of pulpwood and saw timber standing in a given area by measuring the tree heights and diameters. Testing the contestant's ability to improve a timber stand by harvesting select trees because of insect infestation, disease, lack of vigor, or tree conformation is another part of the event.

Foresters must develop good math skills. Many measurements and calculations are required to manage timber land efficiently and effectively. The event tests the student's ability to make ocular estimates of trees and calculate the number of acres and placement of boundary lines in a given area. The FFA team must read a compass and decide points and distances during the compass course. Members must identify not only trees but also the insect and diseases that attack the trees. The FFA Forestry Event requires a winning team to spend time studying classroom materials and practicing skills. State winning teams compete at the National Forestry Contest.

Forestry events allow students to display their skills in forestry measurement and management. *Courtesy of Blane Marable, Morgan Co. High School, Madison, GA.*

The Forest Management Proficiency awards encourage the use of forestry practices that conserve and increase the productivity and economic value of land. The award is presented to individual FFA members who have excelled in experiences using the principles and practices of today's forestry industry. To apply, the student must be involved in a forestry related Supervised Agricultural Experience Program (SAEP). The student must maintain a list of forest management skills learned and practiced during the project. The student also must develop an inventory of buildings, equipment, tools, and materials used in the experience program and keep accurate records of income and expenses of forestry products produced, harvested, and sold. By keeping a list of accomplishments such as land preparation, planting, seeding, thinning, and harvesting, the student can show how he or she has improved the land use for forestry products.

Maria won the Forestry Management Proficiency Award in her FFA chapter. Her program in forestry developed from her interest in her agriculture forestry class and part-time job at a tree nursery. Over a four-year period, Maria recorded 856 hours working for the nursery and developed skills in customer service, propagation, planting, and maintenance of trees. Her experience at the nursery gave her the idea to propagate and grow Leyland cypress as Christmas trees. At her home she rented an acre of land from her parents and planted 200 1-gallon Leyland cypress trees that she bought from her employer. The first year she also built a 10' x 15' cold frame and began propagating cuttings. During the second and third years, she pruned, fertilized and kept weeds away from her trees and sold rooted cuttings to local nurseries. The fourth year, Maria had about 50 trees that were large enough to sell. She pruned and sprayed the trees, preparing them for the Christmas season. The first week of December she cut the trees and sold them at the local farmers' market for $25 each. The fifth year she sold 87 trees and replanted the trees with two gallon plants that she had propagated earlier. By her senior year in high school, Maria had an annual net income of over $6,000. Her agriculture class, part-time job, and supervised agriculture experience provided the knowledge, experience, and opportunity to profit from her interest of growing trees.

FIGURE 11–8
The limbs of faster growing trees unite to form a canopy. *Courtesy of Jim Rathert, Missouri Conservation Department.*

the other species of trees. Also, some species just naturally grow faster than other species. The limbs of the faster growing trees eventually form what is called a **canopy** or overstory above the other trees (Figure 11–8). This prevents trees of the same species from growing in competition. Trees that are not tolerant of shade will not grow under the canopy. However, trees that are tolerant to shade continue to grow at a slow pace.

Eventually the trees that are shade tolerant may overtake and suppress the trees that first created the overstory. This is particularly true if the first trees mature and die. A good example of this process is the forests that covered the upper South. The first trees to dominate were the fast-growing pines such as the loblolly, slash, and shortleaf. These species created canopies or overstories that suppressed the species of trees that were not tolerant of shade. However, beneath this canopy, hardwood species such as red oak and white oak grew. These species tolerate shade fairly well and will grow beneath the pine canopy. A characteristic of the southern pine is that it will mature and die at an earlier age than the oaks. When a pine dies, the space in the canopy may be filled with a hardwood such as oak. When the hardwoods create a canopy, little sunlight will penetrate. Because pines are intolerant of shade, they do not grow well under the oaks. The hardwoods eventually become the dominant species. This is called the climax vegetation because it is unlikely that other species will overtake the hardwoods (Figure 11–9). Succession is a very slow process that may take 200 years to reach the climax vegetation stage. When the Europeans explored this region of the continent they found a mixture of pine and hardwoods. The farther south they went, the more pine they found because the milder climate and sandy soil of the coastal plains favored the pines.

stop @ end?

FIGURE 11–9
Hardwoods eventually become the climax vegetation. *Courtesy of James Strawser, The University of Georgia.*

The Forest Ecosystem

Forests play an important part in our lives because they are a part of our ecosystem. An ecosystem is an area where all the living and nonliving features of the area depend on each other. The forests of the world are instrumental in the support of life on the planet. Plants use carbon dioxide and give off oxygen in their life processes. Because animals use oxygen and give off carbon dioxide, plants support all animal life, including humans. Forests not only provide oxygen, but also aid in removing air pollution.

Young, rapidly growing trees produce more oxygen than older trees (Figure 11–10). As a tree produces new wood, carbon dioxide is removed from the air and oxygen is added. According to The American Forest and Paper Association, in the period of one year, an acre of healthy trees may produce 4,000 pounds of wood, use 5,889 pounds of carbon dioxide, and give off 4,280 pounds of oxygen. Since we have almost 500 million acres of trees, this represents a lot of oxygen that is put in the atmosphere.

FIGURE 11–10
Young, rapidly growing trees produce more oxygen than old trees. *Courtesy of Georgia Pacific.*

FIGURE 11–11
Forest vegetation helps keep streams clean. *Courtesy of James Strawser, The University of Georgia.*

Forests provide us with clean water (Figure 11–11). Roots of trees prevent water from running rapidly down a slope and instead cause the water to soak into the ground, where it accumulates. This recharges streams, lakes, and wells without adding the soil from runoff and helps keep the soil in place on slopes. In doing this, trees prevent lakes and streams from being muddy. Muddy lakes and streams prevent fish and other aquatic life from growing and lower our supply of clean drinking water.

Forests provide habitat for animals and other plants. Many species of plant life live beneath the canopy of trees where they are sheltered from direct sunlight. These plants would not thrive in the open sun. Many species of wildlife inhabit the forests and depend on the trees for their existence. Different types of forests support different types of wildlife. Animals and birds prefer particular types of foliage for food or to make nests in. Without this habitat, certain species would become extinct in a short while. The federal government has established laws and regulations regarding the protection of animal and plant species in the forests. Loggers and timber companies have to abide by the regulations in the production and harvesting of timber (Figure 11–12).

The United States is one of the most beautiful countries in the world. In large part, this is because of our forests. Millions of acres of land are set aside by state and federal government to protect the natural beauty of our forests. These national and state forests preserve a portion of the virgin forests that were here when the first settlers arrived. Today, these areas provide a valuable resource in terms of recreation (Figure 11–13). Hiking, fishing, hunting, and camping are favorite pastimes of many Americans and our forests make these hobbies possible.

One of the greatest contributions our forests make is the large amount of jobs associated with the forestry industry. Billions of board feet of lumber are grown, harvested, and processed each year. This requires a gigantic workforce to care for the trees, remove them from the forest, and convert the logs into finished products. This helps keep the nation's economy sound. It also supplies us with essential products that are used in almost all aspects of our lives.

Wood Fiber Production

One of the largest segments of agriculture is the planting, management, and harvesting of trees. Americans at one time thought that the supply of timber was inexhaustible. Then around the turn of the twentieth century, people began to realize that it was possible to run out of forests to harvest if measures were not taken (Figure 11–14).

Many people also thought that the methods used to harvest timber were wasteful and led to the depletion of the soil. As a result, in 1891 Congress passed the Federal Forest Reserve Act. This legislation gave the president the power to set

FIGURE 11–12
Loggers have to abide by regulations regarding timber harvest. *Courtesy of James Strawser, The University of Georgia.*

FIGURE 11–13
Our forests provide us with areas for recreation. *Courtesy of Vick Christ.*

FIGURE 11–14
Around the turn of the century people realized that the forest would one day be depleted unless actions were taken. *Courtesy of Northwest Forestry Association and Evergreen Foundation.*

aside timberlands to be managed so that people in the future would have forests to use and enjoy. Later, large tracts of land were designated as federal timber lands. The National Forest Service was established to manage these lands. Legislation over the last 100 years has set aside land as being publicly owned. Some land is designated as national parks and is used only for recreational purposes. Other land was set aside as where **wilderness areas** the environment is carefully preserved so that it will remain as close to its original natural state as possible. Still other public timberland is harvested by companies who buy the rights to cut the timber.

Forestry was once looked on as being a natural resource that could be utilized. However, in recent years the timber industry has become an agricultural enterprise. Trees are planted, cultivated, and harvested in much the same manner as other agricultural crops. This means that instead of being a resource that will one day be exhausted, the timber industry can be sustained year after year through the careful planting and cultivation of trees.

FIGURE 11–15
A large portion of the trees harvested in the South go into plywood. *Courtesy of American Plywood Institute.*

For many years, areas clear-cut of all or most of the trees have been seeded or set out with tree seedlings. This allows the forest to grow back in much the same manner that it originally did. Throughout the Northwest, forests have long been regenerated by scattering seeds on newly cut forest areas or by setting out small trees. For a forest to grow from seedlings to harvestable timber may take from 70 to 100 years. For this reason much of the timber industry has moved to the Southeast. In fact, the leading producer of wood is not Washington or Oregon, but Georgia. In this region, the hot humid weather allows the trees to grow in much less time. A carefully managed stand of timber can go from seedlings to harvestable trees in as little as 15 years. It must be remembered, however, that the timber from the Northwest provides a higher quality lumber for the construction industry. A large portion of the trees harvested in the South go for making paper or for plywood core material (Figure 11–15).

Tree Farms

Tree farms are operations where trees are planted in rows and are given care throughout their growth

FIGURE 11–16
Superior tree seedlings are produced in nurseries. *Courtesy of Georgia Pacific.*

period. At the proper time they are all harvested. New trees are then planted and the crop starts over again. There are tree farms in almost all states. The most are located in the South.

The process of tree farming begins in the nursery where superior seedlings are produced for tree plantations (Figure 11–16). Research goes into the selection and production of seeds from high-quality trees that are resistant to diseases and grow rapidly. Many of the seedlings are crosses between two species or strains of trees. For example, a longleaf pine may be crossed with a loblolly pine to produce a tree that is superior to either of the parent trees. This cross, known as a hybrid, benefits from a phenomenon known as heterosis or hybrid vigor. The hybrid grows more rapidly and is generally healthier than the purebred strains of trees.

After the trees are set out in large areas called plantations, they are managed using techniques that interrupt the process of succession. The underbrush may be cleared by using mechanical means, chemicals, or by prescribed burning. The use of fire in prescribed burning is aimed at destroying vegetation that competes with the trees for water and nutrients. The fire is carefully controlled to prevent damage to the trees and to prevent the fire from escaping.

Succession is also controlled by thinning the trees. As the trees grow and begin to compete with each other for sunlight and nutrients, some of the trees are harvested. Usually, the trees that are cut are those that are slowest growing or poorest in quality. The best trees are kept to grow. The trees harvested in the thinning process are used for fence posts, poles, or paper. As the trees grow and mature they are harvested and used for making paper, lumber, and other uses.

Summary

Forestry is a large part of the agricultural industry. Trees are grown and harvested as a crop. If we continue to replenish the forests, we will always have wood to use to make the many products we enjoy. At the same time, we can enjoy the beauty of green forests.

11 CHAPTER REVIEW

Student Learning Activities

1. Take the side of a preservationist or a logger in a controversy over the proposed cutting of a tract of timber. Prepare your points and arguments and debate someone with the opposite view.

2. Make a list of all of the products in your classroom that are made from trees. Do not forget the paper products. Compare your list with others in the class.

③ Select a particular type of tree that is grown for timber. Research the types of products that are made from the trees. What are the characteristics and uses of the lumber from the trees? Report to the class.

④ Visit a building supply store and list all the different types of lumber sold. Identify the species of tree that produced the lumber and the region or country where the trees grew.

True/False

① ___ One of the largest industries in agriculture is forestry.

② ___ There are more trees in the United States today than there were 100 years ago.

③ ___ Wildlife greatly benefits from the forest industry.

④ ___ Plants use oxygen and give off carbon dioxide in their life processes.

⑤ ___ Forests aid in removing air pollution.

⑥ ___ Most tree farms are located in the North.

⑦ ___ Usually the trees that are cut are those that are growing best.

⑧ ___ Succession is controlled by thinning the trees.

⑨ ___ Preservation activists insist that as much of the old growth timber as possible must be preserved.

⑩ ___ Succession is a very fast process.

Multiple Choice

① At one time forestry was considered a
    a. conservation effort
    b. bad investment
    c. natural resource

② The forestry industry stretches from coast to coast and encompasses around
    a. 500 million acres of trees
    b. 100 million acres of trees
    c. 5 million acres of trees

3 Each year there are close to _____ billion trees planted in the United States.
   a. one
   b. two
   c. five

4 The reason certain types of trees dominate a region is because of a natural process called
   a. deciduous
   b. succession
   c. overstory

5 An area where all living and nonliving features depend on each other is called an
   a. ecosystem
   b. oxygen system
   c. industry

6 The legislative act giving the president power to set aside timberlands is called the
   a. American Forest Act
   b. U.S. Forestry Service Act
   c. Federal Forest Reserve Act

7 Young rapidly growing trees produce more _____ than older trees.
   a. carbon dioxide
   b. oxygen
   c. pollution

8 One of the greatest contributions our forests make is the large amount of _____ associated with the forestry industry.
   a. jobs
   b. carbon dioxide
   c. hunting

9 The leading producer of wood is
   a. Washington
   b. Oregon
   c. Georgia

⑩ Operations where trees are planted in rows and are given care throughout their growth period are called
    a. nurseries
    b. loblollies
    c. tree farms

Discussion

① How does wildlife benefit from the forest industry?

② What is the difference between hardwood and softwood timber?

③ What is a canopy?

④ What are the benefits of forests for animals?

⑤ What is the Federal Forest Reserve Act?

⑥ Which state is the leading producer of wood? Why?

⑦ Why is the harvesting of timber not really a threat to nature?

⑧ What is virgin timber?

⑨ What are the benefits of forests for humans?

⑩ Why did the cultivation of trees begin?

CHAPTER 12

Protecting the Environment

Student Objectives

When you have finished studying this chapter, you should be able to:

- Explain why everyone should be concerned with the environment.
- Discuss the importance of agriculture to a growing population.
- Tell how pesticides have the potential to cause harm to the environment.
- Explain how the threat of harm from pesticides is minimized.
- Discuss the difference between point and nonpoint pollution.
- Tell how agriculturists work to prevent water pollution.
- Explain the importance of wetlands.

Key Terms

environment

natural resources

renewable resources

carcinogen

Environmental Protection
Agency (EPA)

hydrological cycle

aquifers

point source

nonpoint sources

wetlands

Everyone should be concerned with protecting the **environment**. There is only a limited amount of space on our planet. Humans continue to multiply and the population grows. This means that there is less space for each person. At one time in history the population was so small that when conditions became bad, people could move to a new place. The places where people can move to a fresh environment are almost gone. Scientists predict that by the year 2050 there will be over 11 billion people in the world. That is about twice the current population. Most of you will be alive at that time and will have to cope with the large population.

As there will be less space for each of us, it becomes increasingly important that we care for the environment. The quality of life will be directly related to the conditions of the environment. Clean water to drink and clean air to breathe are essential. Most of the beauty we enjoy comes from the nature surrounding us (Figure 12–1). Few things are more pleasing than a green forest or a clear stream. The beauty of the surroundings will have to be protected from destruction. The more people

FIGURE 12–1
Most of the beauty we enjoy comes from the nature surrounding us. *Courtesy of Ray Herren, The University of Georgia.*

living in the environment and the more people using our resources, the greater the likelihood that the environment will become polluted. Resources are those things we have available for our use. **Natural resources** occur in nature and they include water, air, mineral deposits, and soil. Usually when these resources are used up, they are gone forever. **Renewable resources** are resources that can be replaced, such as crops. These crops may include trees, flowers, and plants grown for food. In the future we will have to rely more heavily on renewable resources supplied by agriculture.

Not only will there be less space for all of us, but all of the people will need food to eat. Agricultural systems all over the world will have to supply the needs of over 11 billion people—and that number will continue to rise. With all these people needing space in which to live and work, less land will be available to grow food. Agriculturists will have to produce much more food on less land (Figure 12–2). Because of this, farming practices will have to become more intense, which means getting more production out of fewer acres of land. This will be a challenge to produce enough food and fiber to feed and cloth the world's population without wearing out the land or polluting the environment.

Agriculture always has and always will have an effect on the environment. In order to grow crops, forests have to be cleared, and less room is available for wildlife. To produce crops efficiently, fertilizers and pesticides have to be applied. The constant tilling of the soil can cause erosion (Figure 12–3). Animals that are raised in confinement produce a lot of manure that has to go somewhere. The answer cannot be the elimination of agriculture. We must have food, shelter, and clothing. The answer is that we have to design our agricultural production methods so that they will protect the environment.

FIGURE 12–2
With a huge and growing population, agriculturists will have to produce more food on less land. *Courtesy of James Strawser, The University of Georgia.*

FIGURE 12–3
The constant tillage of the soil can cause erosion. *Courtesy of James Strawser, The University of Georgia.*

Pesticides

Pesticides are substances that kill pests. Each year about 570 million pounds of pesticides are used to aid in the control of insects, weeds, and fungi. This may seem like a lot of chemicals, but you must realize that there are over 300 million acres of cropland in the United States. People have developed negative attitudes over the use of pesticides. Some view their use as being terribly destructive to the environment.

In the past, there have been problems with pesticides. Years ago a very popular insecticide was a chemical called DDT. This pesticide was very effective in killing insects and solved a lot of problems all over the world. However, DDT was very residual. This means that it stayed in the soil for a long time before it broke down. This caused problems with the food chain. Birds that ate insects affected by the insecticide retained DDT in their bodies. Water birds might eat fish that had consumed insects or other aquatic life affected by DDT. The DDT built up each time the birds ingested the chemical. Problems developed in the shells of the eggs the birds laid. The shells were so thin that the eggs would not hatch. This interrupted the life cycle of the birds. DDT also turned up in other forms of life. The chemical was passed up the food chain and all of the animals in the chain were affected. Because of these problems, DDT and chemicals like it have been banned. Modern pesticides break down very rapidly and pose much less danger than the older pesticides.

Sometimes, problems with agricultural chemicals are blown out of proportion. For example, in 1989 the media began a series of stories on the chemical daminozide, known by its trade name, Alar. This chemical acts as a growth regulator that is used on apples to prevent the apples from falling off the tree too early. This helps the apples to grow

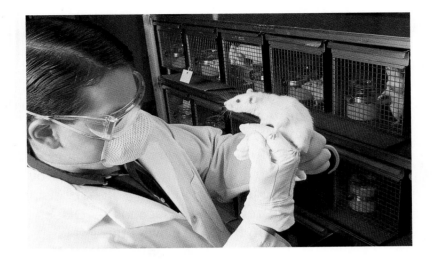

FIGURE 12–4
The chemical Alar was reported to be a carcinogen to laboratory animals. *Courtesy of Dow/ Elanco-French Studios.*

into a better shape. Also it prolongs the storage life. According to the reports, the chemical proved to be a **carcinogen** (a substance that causes cancer) when fed to laboratory rats (Figure 12–4). As a result, near panic ensued among consumers. Apples and apple products were taken off the market. Tons of apples were dumped as garbage and the fruit was banned from school cafeterias. The apple industry suffered tremendous losses. The money earned from the sale of apples fell by more than $100 million. In addition, millions of dollars of income were lost from all the support services of the apple industry. Many apple producers reached the brink of bankruptcy. This example points to the concern of the public about the safety of food.

Further investigation showed little justification for the panic. In the study, laboratory rats had been given enormously high amounts of Alar before they developed tumors. At much lower rates the chemical had no effect at all. According to the American Council on Science and Health, a person would have to eat the equivalent of 28,000 pounds of apples per day for 10 years to equal the amount of Alar fed to laboratory animals before they developed tumors. The amount of Alar that was on the

FIGURE 12–5
A person could not eat enough apples in a lifetime to receive the amount of Alar given to the lab animals. *Courtesy of James Strawser, The University of Georgia.*

apples was almost unmeasurable. In no way could a human eat enough apples in a lifetime to consume enough Alar to cause problems (Figure 12–5). Yet in spite of this fact, the chemical was pulled off the market.

The use of pesticides are necessary if we are to provide the population with high-quality food. Without them both the quantity and the quality of food would be lower. Produce that is damaged by insects or disease will be of such low quality that consumers will be reluctant to accept it. Disease causes bad spots on fruits and vegetables. Fruits and vegetables that have been partially eaten by insects will not only look bad and taste bad but will also spoil more easily.

Used properly, modern pesticides are safe to use. The United States Department of Agriculture (USDA) and the **Environmental Protection Agency (EPA)** have strict regulations regarding the development and use of pesticides. Before a pesticide can be released for use, it must undergo years of rigorous testing to ensure it is safe. It must not only pose no threat to humans but must not be harmful to the environment. Modern pesticides are not only necessary but safe if used properly (Figure 12–6).

FIGURE 12–6
Pesticides are necessary to provide wholesome food. This fruit is not edible because of pest damage. *Courtesy of James Strawser, The University of Georgia.*

Water Pollution

One of the most precious natural resources we have is our water. This compound is essential for life. Without it, humans would die within a few days. We depend on it to sustain the lives of the plants and animals we grow for our food supply (Figure 12–7). Most of our industries require large amounts of water, and so does much of our recreation. Fishing, boating, and swimming require a huge amount of clean water. We have only a certain amount of water. Water is used and recycled over and over again through a process known as the **hydrological cycle** (Figure 12–8). Water evaporates, condenses, and returns to the earth as rainfall.

Rainfall replenishes two types of water storage—surface water and groundwater. Surface water is the water that is in streams, lakes, and the ocean. When rain falls, the amount that is not absorbed into the ground flows downhill into streams. Small streams flow into larger streams and the large streams eventually flow into the oceans. A few years ago, surface waters in this country faced serious pollution problems. Many streams and lakes were so polluted that very little life could exist there.

The other type of water is called groundwater. Groundwater is the water that is stored under the surface of the ground. When rain hits the ground, some runs off and some is absorbed into the ground and accumulates in storage areas called **aquifers**. Aquifers are formations in sand, rock, or gravel that hold water (Figure 12–9). Wells are drilled into aquifers to retrieve water. A large portion of our drinking water comes from aquifers.

Groundwater pollution can be serious. Fortunately, as water travels through the ground, it is filtered and most pollutants do not reach the groundwater. However, where accidental chemical spills occur, problems can happen. However, the

FIGURE 12–7
We depend on clean water to sustain life. *Courtesy of James Strawser, The University of Georgia.*

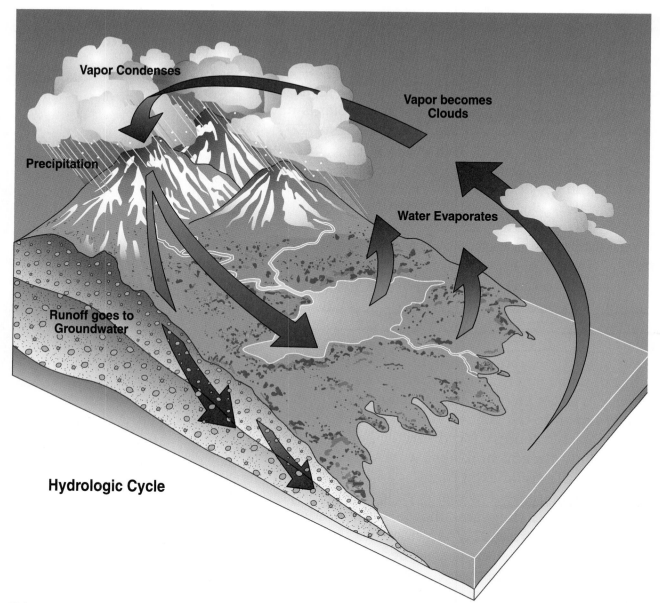

FIGURE 12–8 Through the hydrological cycle, water is constantly recycled.

EPA has determined that fewer than 1 percent of the rural domestic and community water systems contain unacceptable levels of pesticides.

Areas that have shallow aquifers or coarse, sandy soil are in more danger of pollution. The deeper the

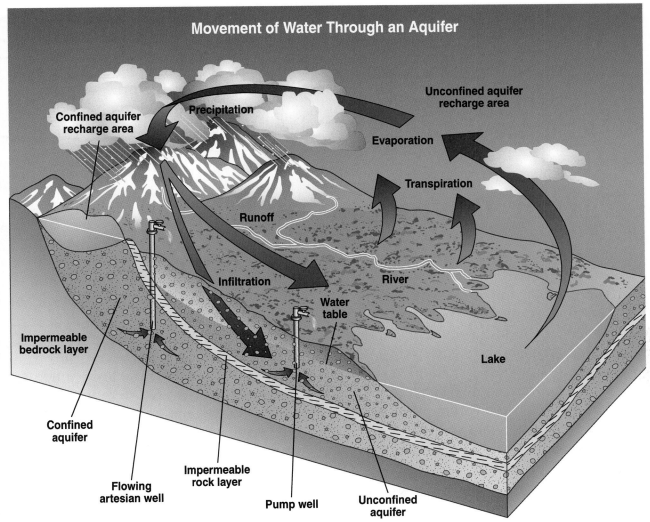

Movement of Water Through an Aquifer

FIGURE 12–9 Aquifers are formations in sand and rock that hold water.

water supply and the heavier the clay content of the soil, the safer the water is. This is because the depth and the clay slow down the movement of the water. This gives more time for the water to be filtered. New regulations governing the use of pesticides have lessened the likelihood of pollution of the groundwater.

In 1970, President Nixon signed into law the National Environmental Policy Act, which led to the establishment of the Environmental Protection Agency (EPA). This agency was charged by Congress to "protect the nation's land, air, and water systems." Policies, laws, and regulations have been and continue to be established to prevent pollution and to clean up pollution that has already occurred. Since 1970, a lot of progress has been

Career Development Events

In every community across the nation there is a need for improvement. The FFA program encourages students to identify community needs and take action. Through the years, community programs have developed active, experienced, and knowledgeable leaders by educating and involving students. Cities, towns, and communities across America have benefited from these projects. Projects have created and expanded community services and improved the quality of life. Community development is one of the major areas addressed in a local FFA chapter program of activities. To make a meaningful impact upon the community, the whole chapter should be involved in the planning and action. The first step is to select a chairperson and committee. The chairperson will lead the chapter in the planning, initiation, and public relations for the effort. The committee will identify needs through questionnaires, attending public meetings, or surveys of the community. After collecting the information, the committee will assess the information and identify a community project. Identification of resource persons and community groups for support of the activity is important. The involvement of groups such as FFA Alumni, Chamber of Commerce, and civic clubs can provide not only financial support, materials, and equipment but offer expertise in needed areas. Cooperative efforts with other community groups will increase the success of the project. A plan of action is necessary to organize the phases of the project and its completion. A step-by-step project plan, work schedule, budget and available resources should be included in the plan of action. After developing the plan of action it is time to start the project. Involving all FFA members and interested community groups will result in a successfully completed project. Publication of community development

made in cleaning up the air and water. Many bodies of water that were once considered dead now have been brought back to life by cleaning up the water and stopping pollution.

Pollution comes about as a result of contaminants entering the water in two main ways: point and nonpoint sources. A **point source** means

activities through newspapers, radio, and television will increase community awareness and support. After the project is completed, a community-wide meeting, such as an FFA banquet, should recognize those groups and individuals who participated. You can make a real contribution to your community!

In competing in the Building Our American Communities (BOAC) Award, students help to improve the environment. *Courtesy of Blane Marable, Morgan Co. High School, Madison, GA.*

pollution from a specific place. For instance, a factory that dumps waste in a river is a point source of pollution because the contaminants can be traced to a single point. The EPA has made tremendous strides in halting point source pollution. Industries have been required by law to stop dumping or allowing contaminants to enter waterways. Water used in industry and returned to streams or lakes is carefully monitored to ensure that it contains no polluting substances (Figure 12–10).

Pollution from **nonpoint sources** is more difficult to deal with. Nonpoint means that the pollution does not come from a single point but from a wide area that is difficult to identify. The pollution is usually from a number of sources that collectively affect the environment. Agriculture is the largest origin of nonpoint pollution in the United States. This is due in part to the enormous size of the agricultural industry, where millions of acres of land are in production. Each year, this land receives treatment in the form of plowing, cultivating, storage of animal wastes, and the application of pesticides. As a result, agricultural pollution comes from several sources: fertilizers, pesticides, animal wastes, and

b

FIGURE 12–10
The EPA requires that water returned to a stream be clean. *Courtesy of James Strawser, The University of Georgia.*

soil erosion. Recently, laws have been passed to help protect the environment from these pollutants.

[Stop!]

Soil Erosion

Our largest environmental problem stemming from agriculture is soil erosion (Figure 12–11). Erosion is the wearing away of the soil by wind or water. This problem is twofold: the loss of topsoil and the pollution of surface water by the displaced soil.

Research indicates that the sediments from soil cause around $6 billion in damage each year. Almost all improved land in the United States has a system of drainage ditches that carries off excess water. Eroded soil clogs these waterways and has to be periodically removed. Similarly, navigable waterways can become so clogged that they have to be dredged out in order for the vessels to pass through. This is a tremendously expensive process that takes a lot of time and resources.

Another serious problem caused by sediment pollution is the loss of fish and other aquatic life in lakes and streams. Some species cannot tolerate

FIGURE 12–11
Soil erosion is our most serious pollution problem stemming from agriculture. *Courtesy of USDA Natural Resources Conservation Service.*

FIGURE 12–12
Some species of aquatic life cannot tolerate muddy water and die from the lack of oxygen. *Courtesy of USDA Natural Resources Conservation Service.*

muddy water and die from lack of oxygen or lack of clean water (Figure 12–12).

As the agricultural industry has developed, methods of protecting the soil from erosion have developed also. Intensive efforts began back in the 1930s with the creation of the Soil Conservation Service (SCS). The mission of this agency was to help prevent the loss of soil and to stop pollution caused by erosion. The SCS has made tremendous strides in the control of erosion. The name was changed in the early 1990s to the Natural Resources Conservation Service (NRCS). Modern efforts are aimed at keeping as much vegetation on the ground as possible. Plants hold the soil together by creating a support system for the soil particles. Roots penetrate the soil and through their fibrous structure hold the soil in place. Many crops are now grown with a bare minimum of plowing (Figure 12–13). This not only allows more plant roots and plant cover to hold the soil but also slows the flow of water over the land. In some areas, crop residues

FIGURE 12–13
In some areas the soil is tilled as little as possible to prevent soil erosion. *Courtesy of James Strawser, The University of Georgia.*

were plowed under but now are left until spring. The new crop is planted in the stubble left from the previous crop.

New varieties of grasses and other plant covers are also being developed. Bare hills are planted in grasses that grow quickly, have an expansive root system, and do well on eroded soils.

Nitrate Pollution

Water is also polluted by agricultural production through the use of commercial fertilizers and animal wastes. Animals that are raised in confinement produce a lot of manure in a concentrated area (Figure 12–14). This material is generally used on fields as a fertilizer. Within both fertilizers and manures are plant nutrients called nitrates. In surface water, a major problem is that nitrates cause plant life in streams and lakes to flourish. This may not seem like a serious problem, but too much plant matter in the water can cause trouble. Clogged waterways are difficult for boats to maneuver through and interfere with recreational

FIGURE 12–14
Animals in a confinement operation produce a lot of manure. Most of it is spread on fields as fertilizer. *Courtesy of James Strawser, The University of Georgia.*

activities. Also, when the plant material dies and decays, oxygen is taken from the water. When a large amount of vegetation decays, much of the oxygen can be removed from the water and the fish may die.

Two practices can help control the problem. The first is to control the runoff from rainwater. Diversion ditches are cut into the area surrounding the livestock facility to prevent runoff containing waste materials from getting into streams. The drainage ditches divert the runoff into holding areas where the risk from runoff is at a minimum.

The second practice is to use a lagoon as a holding tank for the manure. Basically a lagoon is a pond of water into which the waste is directed. Here the material is decomposed by anaerobic bacteria (Figure 12–15). These bacteria live beneath the surface of the water in the lagoon and thrive without oxygen. Anaerobic bacteria break down the waste material and convert the nitrogen into a gaseous

FIGURE 12–15
A lagoon is a pond of water into which waste material like manure is directed. The material decomposes in the water. *Courtesy of James Strawser, The University of Georgia.*

state. At the proper stage of decomposition, the material is pumped from the lagoon and spread on the fields.

Wetlands

Many areas in our country are known as **wetlands**. A wetland is an area where the ground is naturally saturated for much of the year. These areas were once thought to be wastelands that served no good purpose and that should be drained to be productive. It is estimated that over half of the wetlands of the United States have been destroyed for construction or for agricultural purposes.

Scientists now know that wetlands play several essential roles in the ecosystem of an area (Figure 12–16). These swampy areas help to filter water before it reaches open areas that store surface water. Earlier in this chapter, problems with chemicals, sediments, and other pollutants in runoff water were discussed. Wetlands catch runoff water before it reaches deeper bodies of surface water. The vegetation removes nutrients and other pollutants from

FIGURE 12–16
Scientists now know that wetlands play several essential roles in the ecosystem of an area. *Courtesy of James Strawser, The University of Georgia.*

the water and as a result the open areas of surface water are much cleaner. Wetlands also act as a sponge or storage area that collects and absorbs flood water before it can damage crops, homes, or other property.

Wetlands are critical to the life cycle of many different plants and animals. They may provide water for drinking, food, or shelter for a variety of animals in the area. Also, some plant species will only grow in areas where the ground remains saturated for most of the growing season (Figure 12–17). If wetlands are destroyed, many of these plants may be lost forever.

The EPA has established regulations that protect wetlands by restricting their use. Incentives are offered to landowners to promote the preservation of wetland areas. Permits must be obtained from the government when wetlands are to be disturbed. Through these measures valuable wetlands will be protected and in turn the whole environment will be enhanced.

FIGURE 12–17
Some species of plants will only grow in wetlands. *Courtesy of James Strawser, The University of Georgia.*

Summary

Agriculturists will continue to help protect the environment. It is in their best interest and in the public interest. We must have food and fiber. This can only come about through agriculture. Scientists and other concerned individuals will continue to find new and better ways of producing food and fiber and of protecting the environment.

12 CHAPTER REVIEW

Student Learning Activities

① Survey your community and decide if the environment in your area is being cared for. If so, list the measures taken to preserve the environment. If you feel there is room for improvement, outline steps that can be taken to solve the problems.

② Locate an area where there has been soil erosion. Outline a plan that can help prevent more loss of the soil. Be sure to list the major factors that contributed to the erosion.

③ Go to the library and locate a story of a pollution problem that was solved. Report to the class on your findings.

④ Talk to several elderly people who have lived for many years in your area. Get their perspective on how the environment has changed. Do they think the environment is better or worse? Report to the class.

True/False

① ___ The deeper the water supply and the heavier the clay content of the soil, the more likely the water will be polluted.

2 ___ The National Environmental Policy Act led to the establishment of the Environmental Protection Agency.

3 ___ Since 1970, a lot of progress has been made in cleaning the air and water.

4 ___ Erosion is the wearing away of soil by wind or water.

5 ___ Research indicates that the sediments from soil cause approximately $6 billion in damage per year.

6 ___ Diversion ditches are ponds of water where waste is directed.

7 ___ Over half of the wetlands in the United States have been destroyed for construction or agricultural purposes.

8 ___ The Soil Conservation Service (now NRCS), is concerned with minimizing pesticides in soil.

9 ___ All crops are now grown with extensive plowing.

10 ___ Nitrates cause plant life in streams and lakes to flourish.

Multiple Choice

1 One of the most precious natural resources we have is
   a. crops
   b. soil
   c. trees

2 Rainfall replenishes two types of water storage: groundwater and
   a. surface water
   b. lagoons
   c. stored water

3 The water in streams, lakes, and oceans is
   a. groundwater
   b. surface water
   c. polluted water

4 Formations in sand, rock, or gravel that hold water are
   a. wells
   b. wetlands
   c. aquifers

5 Pollution comes about as a result of contaminants entering the water in two main ways:
   a. point and nonpoint sources
   b. rain and snow
   c. fish and birds

6 The largest origin of nonpoint pollution in the United States is
   a. rain
   b. industry
   c. agriculture

7 Our largest environmental problem stemming from agriculture is that of
   a. soil erosion
   b. dairy farming
   c. excess pesticides

8 Permits must be obtained from the government when these are disturbed:
   a. wells
   b. wetlands
   c. aquifers

9 Years of rigorous testing are required to ensure the safety of
   a. crops
   b. wells
   c. pesticides

10 Resources that can be replaced are
   a. renewable
   b. residual
   c. hydrological

Discussion

1 What are natural resources?

2 What are renewable resources?

3 Why must farming practices become more intense?

4 What is the difference between modern and older pesticides?

5 What are some uses for water?

6 Where does agricultural pollution come from?

7 What is soil erosion?

8 Why is sediment pollution such a large problem?

9 Why is it better that crops are grown with a bare minimum of plowing?

10 Why are wetlands so important?

CHAPTER 13

Organic Agriculture

Student Objectives

When you have finished studying this chapter, you should be able to:

- Discuss the meaning of the term organically raised product.
- Explain the concerns some people have about our food supply.
- List ways organic producers fertilize their crops.
- List ways organic producers control insect pests.
- Analyze the controversial issues surrounding organically raised products.

Key Terms

organic	composting	crop rotation
organic fertilizers	humus	border plants
commercial fertilizers	soil amendments	predator insects
cover crop	fish slurry	natural toxins

We live in a very affluent society in which we buy almost everything that we consume and food is no exception. Many years ago, just about everyone grew at least some of the food they ate. If people did not farm for a living, they at least grew a garden and kept cows for milk and hens for eggs. Because they grew most of the food eaten, people knew where their food came from and how it was produced, and so they felt the food supply was safe.

In a society in which few people grow food, most are not aware of where their food comes from, or how it is produced. Food is produced in huge quantities and most often the production takes place many miles from where it is consumed. The efficiency of American producers is due to the use of technology, which has led to the use of chemical fertilizers and pesticides. This has allowed us to produce enough food to feed all of our citizens and much of the world's people.

Advances in laboratory techniques have allowed scientists to identify minute quantities of pesticide and chemical residues on foods. Just a few years ago, these small amounts of residue could not be detected. Regulations by the USDA (U.S. Department of Agriculture) specify the maximum amount of residues allowed on produce sold in this country. Many people feel that any chemical residue on food is too much and that any food produced using chemicals such as pesticides and commercial fertilizer is not healthy (Figure 13–1). They reason that for food to be the most healthy and wholesome, it should be produced using only "natural" inputs.

These concerns have led to a new agricultural industry call **organic** agriculture. In this system food and fiber are produced using only materials that are "organic" in origin. Organic materials are generally considered to be composed of substances

FIGURE 13–1
Many people consider any food that has been sprayed by pesticides to be unhealthy. *Courtesy of Getty Images.*

that occur only in nature and did not result from a manufacturing process.

The meaning of "organically produced foods" is so broad that it is almost impossible to define. The USDA allows other entities to certify products as produced organically. There are several organizations that set rules and regulations for labeling food and fiber as organically produced. For example, an organization called California Certified Organic Farmers sets the regulations for organically grown products in California. They determine whether or not the producers and products meet the qualifications set forth by their organization. If the qualifications are met, the product may be labeled as "organically grown." Many other states have similar organizations.

Organic farming has been growing steadily for over 30 years (Figure 13–2). Today there are almost 1½ million acres of organically produced crops in the United States. This acreage increases each year.

The Production Process

Methods used by organic producers are a lot different from those used by conventional producers.

FIGURE 13–2
Markets for organic foods have grown steadily.

A majority of all the food and fiber products produced in the United States are grown using **commercial fertilizers** and pest controls. Also, much of the beef is grown using growth stimulants or hormones to improve the growth efficiency of the animals. Organic producers use a different production practice. They use natural methods of fertilizing and pest control.

Fertilizers

Producers of organically grown foods and fiber use a variety of methods to provide plants with nutrients. The idea is to use only materials occurring in nature and not to use fertilizers made purely from chemicals. One of the basic methods is to use **cover crop** plants such as annual winter grasses or legumes that are planted each year when crops are harvested. The cover crops grow during the winter and early spring until the time when new crops are to be planted. They are then turned under before the next crop is planted to provide organic material for the new plants (Figure 13–3). The cover crop decays and through the natural process of decomposition releases nutrients into the soil.

FIGURE 13–3
Organic producers often plant cover crops to add organic material to the soil. *Courtesy of Getty Images.*

In addition, cover plants such as vetch and clover are legumes that have nodules on the roots that host nitrogen-fixing bacteria. These bacteria take nitrogen from the air and convert it to a form of nitrogen useable by plants. The process of nitrogen fixation greatly improves the producing ability of the soil.

Another method used by organic producers is that of **composting**. Materials such as leaves, grass clippings, and various other plant parts are put in a bin designed to speed the decomposition process. As plant material decays it is converted into a substance called **humus**. Humus contains nitrogen and also aids in the water- and nutrient-holding capacity of the soil. Other organic materials such as peat moss or decayed sawdust (called **soil amendments**) are added to the soil. Other amendments such as lime or gypsum may be used to adjust the pH level of the soil to suit the particular crop being grown.

There are several sources where producers of organically grown products can obtain materials for use as fertilizers (Figure 13–4). One is the fishing industry. When fish are harvested and dressed, a large portion of the fish carcass remains after the

FIGURE 13–4
Organic materials are composted to produce natural fertilizer. *Courtesy of Getty Images.*

edible portions are removed. This combination of heads, skin, entrails, fins, and bones is ground and blended into a slurry that is used as a nutrient-rich fertilizer. **Fish slurry** is high in nitrogen content and also contains phosphorous and potassium. These three nutrients are all major nutrients required by plants.

Perhaps the most important source of organic fertilizer is animal manure. All areas of the animal industry produce millions of tons of fecal matter each year that has to be disposed of properly. Because manure is very high in nutrient content it is a valuable fertilizer. The dual purpose of getting rid of waste and fertilizing crops is then achieved. Unless the proper steps are taken, major problems may exist with using manure. One problem is that manure causes quite an odor problem when it is spread onto a field near where people live.

Producers who grow crops near a residential area and fertilize with raw manure can come under heavy criticism from people living in the area. A second problem with using manure is that the fecal matter contains high levels of *E. coli* and other forms of bacteria that may be harmful to humans.

A third problem with using manure is that most raw manure contains a considerable amount of weed seed. When grain is grown almost always some amount of weeds will grow up and mature in the field alongside the crop. When the crop is harvested, the seeds from the weeds are also harvested. Many weed seeds are too small to separate out when going through the processing with the grain. Because most of the weed seeds are not harmful to livestock, they are left in the grain as part of the processed livestock feed. The weed seeds are then ingested by the animal, go through the animal's digestive system, and are passed out with the feces. When the feces are used as manure fertilizer, the weed seeds are planted and may germinate in the soil along with the crop.

FIGURE 13–5
If manure is properly composted, problems with odor, weeds, and bacteria can be eliminated.
Courtesy of Getty Images.

Using proper steps in the process of composting can eliminate all three of these problems (Figure 13–5). When the manure is composted it is piled together and allowed to age for at least two months prior to use. If properly composted, internal temperature of the manure should reach 130 to 140 degrees F. This degree of heat should be enough to kill most harmful bacteria and weed seed. In addition, the composting process helps ease problems with odor.

(Stop!)

Insect Control

One of the greatest problems facing organic producers is that of controlling insects. When plants are growing and weather conditions permit, populations of insects can build up rapidly and destroy a crop very quickly (Figure 13–6). Conventional producers use a variety of chemical weapons to control insects. However, organic producers are against using insecticides to kill pests and use more natural methods such as **crop rotation** to help in insect control.

The principle behind crop rotation as an insect control method is that many insect pests attack only particular plants. By planting fields with different crops each year, populations of insects have less chance of growing because the plants they feed on are not grown two years in a row. For example, horn worms may attack tomatoes but may not bother squash. Planting squash in a field that had tomatoes the previous year deprives the horn worms of food. This helps break up the insects' life cycle.

Another method is planting **border plants** around the outside of the field where the crop is grown. These borders serve to provide an alternative crop for the insects. The idea is that the insects attack the plants around the border and not the crop. Border plants are usually perennials that do

FIGURE 13–6
Populations of insects can build rapidly and destroy crops. *Courtesy of Getty Images.*

FIGURE 13–7
Predator insects such as the lady beetle can help control harmful insects. *Courtesy of Getty Images.*

not have to be planted each year and must be plants that insects prefer to the crop plants.

One of the most effective weapons organic producers have in controlling insects is the use of **predator insects**. In nature many insects live by eating other insects. There is a high population of predator insects such as praying mantises, lacewings, lady beetles, and parasitic wasps (Figure 13–7). Many of the predator insects are grown commercially for the purpose of selling to organic producers. Modern garden supply stores offer predatory insects for sale through their retail catalogs. As soon as insect pests appear, the organic producer releases large numbers of the predatory insects to eat the harmful insects.

Specific predators are used for specific pests. For example, lacewings feed only on thrips and aphids, so if these insects are a problem the producer introduces a large number of lacewings. The larvae of parasite wasps eat a variety of insects and are used as a method of killing more than one type of insect pest. The wasp lays eggs on or in the harmful insect and, on hatching, the larvae feed on the insect.

Organic Animal Agriculture

Although most organic production is centered around plants, animals are also grown and sold organically. Most often, livestock products labeled as organically produced applies to animals that have been raised without pesticides to control for internal and external parasites. Also the animals must not have been exposed to the use of growth stimulants or hormones. The labeling also extends to the use of irradiation for preserving meats. Although the USDA allows this method of preservation, meats treated with radiation may not be labeled as organically produced. Animals are generally raised in open-air,

FIGURE 13–8
To be certified as organically produced, animals generally have to be raised in open-air, noncrowded conditions. *Courtesy of Getty Images.*

noncrowded conditions, which include the humane treatment of the animals (Figure 13–8).

Sometimes the certification may require that the animal be fed organically produced feed and raised by other strict nutritional requirements. For example, the California Certified Organic Farmers prohibit the following:

1. Nonorganic feed
2. Feed containing plastic pellets, urea, synthetic preservatives, or other synthetic materials
3. Feed containing antibiotics, hormones, drugs, or synthetic growth promoters
4. Intentional refeeding of manure
5. Maintaining animals on poorly balanced diets, including raising anemic animals
6. Feed containing animal by-products for ruminants
7. Feed that has been solvent extracted

Only products from animals that meet all the production requirements may be labeled as organically produced. This includes products such as wool and hair, meat, milk and milk products, eggs, and any other animal-derived product.

Criticisms of Organic Production

Since the concept of organic production began, controversy has surrounded the issue. Critics maintain that the public is misinformed about the safety and nutritional value of organically grown products. They point out that these products are much more expensive than conventionally produced food and fiber and are generally of poorer quality. Critics point out that a lot of research has been done comparing the safety and nutritional value of organic foods to conventionally grown foods. To date, no definite evidence has shown that organically produced foods are safer or more nutritious than conventionally produced food (Figure 13–9).

In fact, there are several disadvantages. One of the main reasons for organically grown food is to avoid the toxins from the trace amounts of insecticides that could possibly be on regularly produced fruits and vegetables. Plants have a self-defense mechanism that produces **natural toxins** to help repel insects. If attacked by pests, the plants release these substances because of the stress caused by

FIGURE 13–9
Research has failed to show that organically produced foods are more nutritious or healthier than conventionally produced foods.

the insects or disease. These natural toxins can be many times the amount produced by plants that were protected by pesticides.

It is difficult to produce fruits and vegetables that have the quality we are used to without the use of chemical pesticides. Natural methods of controlling pests can be effective, but when used alone cannot offer the protection that will ensure that the food does not contain insect damage. This damage on fruits and vegetables provides a natural harbor for bacteria and other organisms that cause the food to spoil (Figure 13–10). Most scientists agree that the greatest danger we have from our food supply is that of food spoilage and the ingestion of bacteria, not the presence of pesticides.

Another problem with organically produced food is that often manure is used as fertilizers. Unless manure has been treated, it contains bacteria that can be harmful to humans. Critics point out that contact with these bacteria greatly offsets any advantage offered by using manure. Also, they point out that nitrogen, phosphorus, and potassium, which are the major nutrients needed by plants, are the same whether they come from organic or inorganic sources.

Critics argue that organic production requires more land and more labor to produce the same amount of products grown using conventional means. In fact, some scholars argue that intensive farming methods are much more "environmentally friendly" than organic production. Projections are that around the middle of the twenty-first century, the world's population will double. This means that the world's production of food and fiber will have to double through either of two methods. One is to clear new land and increase the number of acres under cultivation. The second is to grow more food and fiber on the land currently under production. The concern raised over organic production is that

FIGURE 13–10
Insect damage on fruits can provide a harbor for bacteria and other harmful organisms. *Courtesy of University of California.*

FIGURE 13–11
Critics contend that not enough food can be produced using organic means to feed the world's population.

its practice will necessitate the clearing of more land. As more land is cleared, the loss of wildlife habitat will increase. Most of the new land being cleared is through the "slash and burn" methods used in tropical areas. Much of the land cleared is poor quality and highly susceptible to erosion. However, through the use of conventional, more intensive farming, less new land will have to be put into production.

The fact that organic production relies on the use of organic materials is also a weakness of the system, especially if this were the main method of producing our food and fiber. If all the animal manure and all the beneficial organic materials available were used, there would not begin to be enough to fertilize all the cropland. Obviously, fertilizer inputs will be needed, and the argument is made that organic sources alone can supply only a small fraction of what is needed (Figure 13–11).

Summary

Despite the controversy over the issue, the production and consumption of organically grown products continues to grow. This relatively new branch of the agricultural industry is here to stay and will receive increasing attention in the years to come. As with all other aspects of the market, it is driven by consumer demand and as that demand increases, production will also increase. The challenge will be to keep consumers well informed on research findings related to organically raised products.

CHAPTER REVIEW

Student Learning Activities

1. Go to the grocery store and make a list of all the products you can find that were organically produced.

2. Research and prepare a 5-minute speech giving your position on organic agriculture.

3. Compare organically produced fruit (like an apple) with conventionally grown fruit of the same type and variety. List the differences in cost, appearance, and taste.

True/False

1. ___ Organic agriculture uses only materials that originated from living material.

2. ___ E. coli is one of the problems when using manure as a fertilizer.

3. ___ Controlling insects is not a very big problem for organic producers.

4. ___ When irradiation is used, the meat can still be considered organic.

5. ___ Most scientists agree that the greatest danger we have from our food supply is that of food spoilage and the ingestion of bacteria.

6. ___ If only organic farming is used in the future, there will not be enough organic material to fertilize all the crops planted.

7. ___ Clearing land for organic farming will have no impact on wildlife.

8. ___ Nitrogen-fixing bacteria take nitrogen from the air and convert it to a usable form for the plant.

9. ___ A problem with using animal manure for fertilization is that it contains weed seed.

10. ___ The least effective weapon organic producers have in controlling insects is the use of predator insects.

Multiple Choice

1. Today there are almost _____ acres of organically produced crops in the United States.
   a. 2 million
   b. 1 billion
   c. 1½ million
   d. 500 thousand

2. Which is not an example of an organic fertilizer?
   a. fish slurry
   b. compost
   c. manure
   d. 10-10-10

3. Surrounding a field with a different crop is an example of
   a. composting
   b. border plants
   c. cover crop
   d. edging

4. Which is an organically produced product from an animal?
   a. wool
   b. meat
   c. milk
   d. all of the above

5. At what internal temperature should manure be heated to kill harmful bacteria?
   a. 100 to 120 degrees F
   b. 220 to 230 degrees F
   c. 180 to 190 degrees F
   d. 130 to 140 degrees F

6. What does USDA stand for?
   a. United States Demands Action
   b. United States Department of Agriculture
   c. Use Safety Doing Agriculture
   d. United States Department of Action

7. Most of the land being cleared is with the _____ method.
   a. slash and burn
   b. clear-cutting
   c. seed tree
   d. shelterwood

8. Which is not an example of a cover crop?
   a. legumes
   b. winter grasses
   c. trees
   d. clover

9. Border plants are usually
   a. biennial
   b. annuals
   c. perennials
   d. none of the above

10. Which is an example of a predator insect?
    a. lacewings
    b. lady beetles
    c. parasitic wasps
    d. all of the above

Discussion

1. Discuss why people are in favor of organic agriculture.

2. Discuss the use of crop rotation.

3. Explain how lacewings are helpful predator insects.

4. List five of the seven items California Certified Organic Farmers prohibit when producing organic animals.

5. Discuss some of the criticisms of organic production.

CHAPTER 14

The Livestock Industry

Student Objectives

When you have finished studying this chapter, you should be able to:

- Discuss the extent of the U.S. meat industry.
- Discuss the different phases of the beef industry.
- Explain how pork is produced in the United States.
- Discuss the extent of the sheep industry.
- Describe the types of and uses for horses in this country.
- Describe the small animal industry.

Key Terms

per capita consumption	carcass	ration
beef	crossbreeds	pork
breed	purebred	

The livestock industry in the United States is a tremendously large industry. Animals are raised in all 50 states. They are used for meat, milk, and dairy products (see Chapter 15), clothing (wool and leather), and recreation (pets and horses). For as long as there has been a country, Americans have raised large numbers of livestock. Our country is well suited for growing livestock. Much of the open land in this country that is not suitable for growing row crops is suitable for growing animals (Figure 14–1).

The Meat Industry

The average person in this country consumes almost 200 pounds of red meat and poultry each year. Very few nations in the world even come close to us in the **per capita consumption** of meat (Figure 14–2).

Some health authorities have expressed concern over the consumption of so much meat. Research has shown that a substance called cholesterol in the fat of meat can cause heart problems. Most health experts agree that a moderate intake of meat is necessary for a healthy diet. Meat is among the

FIGURE 14–1
Much of the land in the United States that is not suitable for row cropping is suitable for growing animals. *Courtesy of James Strawser, The University of Georgia.*

FIGURE 14–2
Very few nations in the world eat as much meat as Americans. *Courtesy of James Strawser, The University of Georgia.*

most nutritionally complete foods that humans consume. Foods from animals supply about 88 percent of the vitamin B_{12} in our diets because this nutrient is very difficult to obtain from plant sources. In addition, meats and animal products provide 67 percent of the riboflavin, 65 percent of the protein and phosphorus, 57 percent of the vitamin B_6, 48 percent of the fat, 43 percent of the niacin, 42 percent of the vitamin A, 37 percent of the iron, 36 percent of the thiamin, and 35 percent of the magnesium in our diets. Problems with cholesterol can be avoided by exercise and eating a reasonable amount of meat.

The Beef Industry

Beef is by far the largest segment of the meat industry. In the United States, there are over 800,000 producers of beef. Over three-quarters of all of the money earned from the meat industry comes from beef. The average size of beef herds in the United States is about 100 animals.

In the United States there are over 40 different breeds and many different combinations or crosses of these breeds. A **breed** is a group of animals with a common ancestry and like characteristics (Figure 14–3). The offspring of these animals look like their parents. Livestock producers choose breeds based on several factors. The market where the animals will be sold, the environmental conditions the animals will be produced in, and the personal likes and dislikes of the individual producer may be considered.

Some breeds are large and produce a large **carcass**, some mature at a smaller size and produce a smaller carcass. A carcass is that part of a meat animal remaining after slaughter and the hide, head, feet, and internal organs have been removed. Both sizes have a place in the market and both are produced. Some breeds are adapted to

FIGURE 14–3
A breed is a group of animals with a common ancestry and similar characteristics. *Courtesy of James Strawser, The University of Georgia.*

hot, humid climates and some tolerate the cold and snow better. A producer may like the color pattern or the docile nature of a particular breed and prefer to produce that breed.

Some breeds make excellent mothers, while other breeds grow rapidly and produce high-quality, meaty carcasses. Because of this, some breeds are referred to as sire breeds and some are referred to as dam breeds. A sire is the father and the dam is the mother.

Most animals produced for slaughter are **cross-breeds**. A crossbreed is an animal with parents of different breeds (Figure 14–4). A crossbreeding

FIGURE 14–4
A crossbreed is an animal with parents of different breeds. *Courtesy of Cooperative Extension Service, The University of Georgia.*

program helps producers to take advantage of the good points of both types of animals.

\#5 The beef industry has four major segments: purebred operations, cow-calf operations, stocker operations, and feed lot operations. **Purebred** cattle are produced in the first phase of the industry. The term purebred means that the animal is of a certain breed and has only that breed in its ancestry. Purebred animals are eligible for registration, which certifies that the animal is purebred. The purpose is to produce what is known as seed stock cattle. These cattle are used as the dams and sires of calves that will be grown out for market. These are usually those that are of the sire type or dam type of animal.

The second phase is the cow-calf operations, where the calves are produced. The calves are weaned from their mothers and sold to the next phase of the industry (Figure 14–5). Most of these calves are crossbred animals from purebred parents of different breeds. A large part of this industry is centered in the southern and western states. The mild winters of the South are ideal for calving in the winter. In the West, producers can take advantage of the vast amounts of government lands that are open to grazing for a small fee. Often, cows are left on free range (not fenced in) to have their calves; the calves are then rounded up, weaned, and sold.

Calves are usually sold upon weaning. They are weaned at about six to seven months in age. They generally weigh approximately 500 to 600 pounds. At this age and weight they are ready to begin feeding for market.

The next phase of the industry is stocker operations, a step between the weaning of the calves and the finishing or fattening of the animals prior to slaughter. The stocker purchases the animals from the cow-calf producer and sells them to the feedlot

FIGURE 14–5
The purpose of a cow-calf operation is to produce calves to be grown out for market. *Courtesy of James Strawser, The University of Georgia.*

FIGURE 14–6
Stocker operations provide a step between weaning and the feedlot. *Courtesy of Ronnie Silcox.*

operator. The stocker provides a transition period for the calves between the time they are weaned and before they are put in the feedlot (Figure 14–6). The calves are fed a **ration**, which is the amount of food necessary during a 24-hour period, that ensures that they make sufficient gains to be placed in the feedlot.

It is not uncommon for feedlot owners to also run stocker operations. Such arrangements are economical in that fewer transportation costs are incurred if the two types of operations are close together. The trend in the industry has been away from the stocker industry as new research reveals production methods that allow cows to wean heavier calves.

The feedlot operation is the final phase before the animals are sent to slaughter. Here the animals are fed on a high concentrate ration designed to put on the proper amount of fat cover (Figure 14–7). A concentrate is a feed that is high in grain content. Grain contains a lot of carbohydrates and helps an animal to fatten. A roughage is a feed such as grass or hay that is high in fiber. The producers usually want their animals to be marketed when the cattle reach a sufficient fat cover to allow the animals to grade "low choice."

FIGURE 14–7
Cattle in a feedlot are fed a high concentrate ration. *Courtesy of American Limousin Association.*

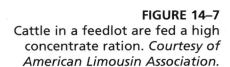

FIGURE 14–8
Pork production and consumption rank second only to beef in the United States.

Many feedlots are situated in the Midwest because this section of the country produces the most grain. It is usually more economical to feed the animals there rather than ship the grain across country. An exception is the state of Texas, which has more feedlots than any other state. When the animals are slaughtered, they are generally around 18 to 24 months in age and can weigh from 800 to 1,500 pounds. This age and size offer consumers the type of beef they prefer.

The Pork Industry

As producers and consumers of **pork**, the United States ranks behind Asia and Europe. Every year Americans produce over 85 million head of hogs. In terms of meat, pork production and consumption rank second only to beef in this country (Figure 14–8). Pork is distributed throughout the country, although some groups, such as Moslems and Jews, do not eat pork for religious reasons.

Much of the pork produced in this country comes from the midwestern states of Iowa, Illinois, Indiana, Minnesota, and Nebraska. These are the states that produce corn, the major grain fed to swine. Also there are large numbers of pigs in the

South, where mild winters help lower the cost of production.

There are not as many breeds of hogs as there are breeds of cattle grown in this country. Breeds are categorized as mother or sire breeds. The mother breeds include Landrace and Yorkshire, which have a large number of pigs per litter and produce lots of milk for their young (Figure 14–9). The sire breeds, such as the Duroc and the Hampshire, characteristically grow rapidly and produce well-muscled, meaty carcasses.

Most of the market hogs produced are crosses of the various breeds. For example, a Yorkshire female mated to a Duroc male provides an excellent cross. The Yorkshire sow provides a large litter and plenty of milk and the Duroc boar sires pigs that are vigorous and fast growing. In addition, crossbred pigs are healthier and do better than purebred pigs.

Two phases of the industry are the farrowing and finishing operations. The two phases can be separate, or they can be operated together. Some producers prefer to raise only feeder pigs (pigs that are

FIGURE 14–9
A Landrace female makes a good mother and is considered a dam breed. *Courtesy of the American Yorkshire Club.*

Career Development Events

If you have an interest in livestock and would like to work in the livestock industry, you could benefit from participating in several FFA career development events. The FFA Livestock Judging event is for teams consisting of four members. Each member competes individually and the final team score is calculated by adding the three highest individual scores. The FFA livestock judging event aids classroom instruction by teaching students practical skills in selecting quality livestock. All livestock producers know that the selection of quality livestock is essential to be competitive and profitable in the industry. Some skills and talents you will need to participate in the event are understanding the breed characteristics of beef, swine, and sheep. The ability to rank classes of animals and give reasons for placement of animals is also required. Team members independently evaluate and place each livestock class. After the judging, students are asked to give reasons for the rationale they used to place the livestock. Points are awarded for each class placing and correct reasons. This FFA event is usually held at the chapter, area or district, state, and national levels.

Another FFA team event related to the livestock industries is the Meats Judging Career Development Event. Like the livestock judging event, teams are made up of four FFA members. As a participant in this event, you will gain valuable knowledge of meat identification and grading. This knowledge will benefit you in the future if seeking a job in the meat packing industry or if you are a consumer of meat. Preparation for this event combines team practice and individual study. The team should practice as often as possible. In most career development events, students identify primal cuts of beef and pork, retail cuts of pork, beef, and lamb, and quality or yield grades of carcasses. By visiting meat processing plants and local grocery meat departments, the identification of wholesale and retail cuts can be reviewed. Each contestant will need to study for the written test relating to meat selection, storage, cookery, nutrition, and safety. Situational problems involving the least-cost formation of a batch of meat products such as hamburger or bologna are given. The student works through procedural questions and the actual determination of the least-cost price. The FFA Meats Judging event develops employment knowledge with practical skill applications preparing students for entry-level employment in the meat packing industry.

Students may compete in individual events in this area through the FFA Proficiency Awards. All FFA members who complete Supervised Agricultural Experience Programs (SAEPs) as part of their agriculture education instruction are eligible for the award. Livestock proficiency awards are available in beef production, dairy production, diversified livestock production, and sheep and swine production. Proficiency Award medals and certificates are provided by the National FFA Foundation. Chapter winners may compete at the state level where the winner earns a cash award and a plaque plus travel expenses to the National FFA Convention. The top FFA member in the nation in each award area is recognized and awarded at the National Convention.

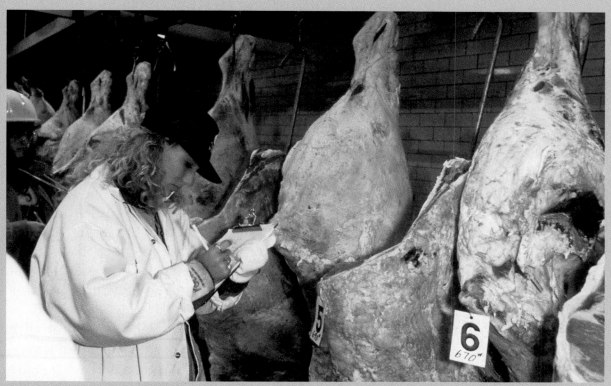

In the Meats Judging Career Development Event, students test their knowledge of meat quality and yield grades as well as the identification of different cuts of meat. *Courtesy of Carol Duval, National FFA Organization.*

FIGURE 14–10
Pigs are generally born in a furrowing crate to protect them from being crushed. *Courtesy of University of Minnesota Agricultural Experiment Station.*

sold shortly after weaning) and some prefer to buy feeder pigs and finish them as their only operation.

Most pigs are farrowed in climate-controlled houses where the mother is kept in a crate to prevent her from injuring the pigs as she lies down (Figure 14–10). Good producers make quite an effort to provide an environment that is clean, dry, and comfortable for both the mother and the piglets. The pigs are generally weaned from the mother at six weeks, although they may be as young as three weeks or as old as eight weeks.

After weaning, the pigs are placed together with pigs of similar age and size in what is called a confinement operation. This means that the hogs are kept in a pen together rather than running loose on a pasture. Sufficient space is allowed for the pigs to be comfortable and to grow at a fast rate (Figure 14–11). The pigs should be finished (reach the proper market weight and condition) at about

FIGURE 14–11
Hogs are grown to market weight on a finishing floor. *Courtesy of James Strawser, The University of Georgia.*

20 weeks. Packers like to buy market hogs that weigh in the range of 220 to 260 pounds. (Stop!)

The Sheep Industry Pick up

Compared to beef and pork, Americans eat relatively little lamb and mutton. Lamb refers to meat from a sheep that is less than a year old; mutton refers to meat from a sheep that is over a year old. In many parts of the world, lamb and mutton are a basic part of the diet. The U.S. per capita consumption of mutton and lamb is only about 2.5 pounds (Figure 14–12). Of this, about 95 percent is lamb and

FIGURE 14–12
Annual consumption of lamb is only about 2.5 pounds per capita in the United States. *Courtesy of James Strawser, The University of Georgia.*

FIGURE 14–13
The Dorset sheep is an example of a medium wool breed of sheep. *Courtesy of James Strawser, The University of Georgia.*

only about 5 percent is mutton. Americans seem to have never developed a taste for the stronger flavored mutton.

\# 6 One advantage of producing lambs for market is that very good quality lambs can be produced on grass without having to be fed a lot of expensive grain. Although an increasing number of lambs are being fed on grain in feedlots, roughages still make up about 90 percent of all the feed consumed by sheep.

The leading states in sheep production are Texas, California, Wyoming, Colorado, South Dakota, Montana, New Mexico, Utah, and Oregon. Sheep can make better use of lower-quality forage than can cattle. For this reason, they can be successfully grazed on poorer grazing lands of the desert areas of the West. In addition, a drier climate helps reduce parasite and disease problems associated with sheep that are grown in the more humid areas.

Sheep breeds are generally grouped according to the type of wool the animals grow. Wool type is broadly classified as fine wool, long wool, medium wool, hair, and fur. The medium wool breeds are used most often to produce lambs for slaughter. Medium wool breeds commonly used to produce slaughter lambs are Suffolk, Hampshire, South Down, and Dorset (Figure 14–13).

The Horse Industry

Humans have used horses for transportation, work, and war as far back as recorded history goes. Almost all civilizations at one time relied on horses or donkeys. In this country, much of our history has been built around power supplied by horses and mules (Figure 14–14).

The number of horses and mules in this country grew until the 1920s, when the car, truck, and tractor caused a sharp decline in their numbers. From that

FIGURE 14–14
Much of our history has been built with the power of horses and mules. *Courtesy of John Deere & Company.*

time until around 1960, the numbers steadily declined. Since the 1960s, the number of both horses and mules has increased dramatically. While some horses and mules in the United States are used for work, most of the animals are used for recreational purposes.

In modern times horses are generally categorized as light horses, draft horses, or ponies. Light horses are animals that weigh 900 to 1,400 pounds. These horses are further divided according to use. Gaited saddle horses and walkers are used primarily for pleasure riding and show. They have different ways in which they walk or trot that seem graceful and flowing. Driving horses are used to pull carriages that are either four or two wheel and are used for recreation or for show.

Stock horses are used all across the country to work cattle or other animals. Race horses are used to compete with each other as a spectator sport.

Draft horses are those breeds that weigh over 1,400 pounds. At one time, these animals provided the power for pulling heavy loads such as wagons, plows, and other agricultural implements. Today, these animals are used in pulling competitions, shows, and parades (Figure 14–15).

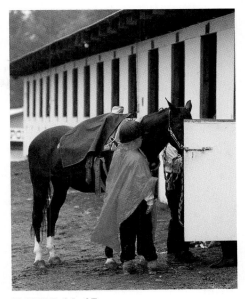

FIGURE 14–15
Today most horses are grown for recreation. *Courtesy of Michael Dzaman.*

FIGURE 14–16
Ponies are breeds of horses that weigh between 500 and 900 pounds at maturity. *Courtesy of Multi-World Champion Shetland Harness Pony.*

Ponies are breeds of horses that weigh 500 to 900 pounds when mature (Figure 14–16). While these are used to pull carriages and for show, the majority of ponies are used as horses for children.

Altogether the horse industry in this country is a $15 billion industry. Surprisingly, horse racing ranks third behind baseball and auto racing among spectator sports in the United States. Each year over 75 million people attend thoroughbred and harness racing. Each year there are around 7,000 horse shows in this country where young people and adults compete in a variety of different events. Many horses are owned as individual saddle horses that are used as recreation and are never raced or shown.

Small Animals

A growing industry in the United States is that of producing small animals. These animals range from rabbits raised for meat to animals such as gerbils and mice raised for pets. Animals such as guinea pigs and rats are raised for research. These animals are used to test everything from drugs to toothpaste.

Rabbit Production

People have raised rabbits for food for hundreds of years. Although most rabbits are produced by part-time growers, there are several large commercial operations in this country. Americans consume around 10 to 13 million pounds of rabbit meat per year. Some of the meat consumed in this country is imported from Europe where the rabbit industry is larger. France is the leading producer.

There are several advantages to raising rabbits. Rabbits are easily raised under almost any climatic condition. Most are raised indoors in cages called hutches that consist of woven wire with boxes for the rabbits to sleep in and to have their young (Figure 14–17).

Many restaurants now offer several dishes that are prepared using rabbit meat. The USDA points out that rabbit meat is one of the most nutritious meats available. Not only is it high in protein and low in fat and cholesterol, but also it is easily digested and is very flavorful. In addition to the meat, rabbits are used for several purposes. The fur is used in making

FIGURE 14–17
Rabbits are raised in cages called hutches. *Courtesy of Calvin Alford, The University of Georgia.*

FIGURE 14–18
In the future, animal agriculture may expand to include animals that are now considered to be exotic. *Courtesy of James Strawser, The University of Georgia.*

coats, hats, liners for boots, and for toys. Scientists use the animals in medical research and other areas. Many animals are sold as pets because of their docile nature, clean habits, and cuddly fur.

Rabbits can be a relatively inexpensive source of meat, but Americans do not consume as much rabbit meat as other peoples of the world. While we may accept the production and slaughter of cattle, hogs, and chickens, the thought of slaughtering cute, cuddly animals such as rabbits is repulsive to many.

Summary

The livestock industry in this country is broad and diverse. Most experts say that in the future it will become even more diverse. In the future production will expand to include some exotic animals (Figure 14–18). Animals such as ostriches, emus, and alligators may be grown for food. This may seem far-fetched, but remember that at one time cattle, horses, and sheep were wild animals before humans tamed them.

 CHAPTER REVIEW

Student Learning Activities

1 Make a list of all the products in your home that come from animals. Determine what type of animal was used for the product. For example, a leather belt was probably from a beef animal and a wool sweater was probably from a sheep.

2 Report on the most widely grown agricultural animal in your state. List the reasons why this animal is popular in your state.

③ Design a plan for growing animals that are not grown in your area. Give your reasons for growing the animals, why your area would be suitable, and how you would market the animals.

④ Go to the Internet and research a breed of livestock. Determine where and when the breed originated. What are the characteristics of the breed?

True/False

① ___ The livestock industry in the United States is a tremendously large industry.

② ___ Our country is not suited for growing livestock.

③ ___ Meat is among the most nutritionally complete foods that humans consume.

④ ___ The average size of the beef herds in the United States is about 50 animals.

⑤ ___ Most feedlots in this country are situated in the South.

⑥ ___ Ostriches, emus, and alligators may be grown for food.

⑦ ___ Calves are usually sold on weaning.

⑧ ___ The United States leads the world in the production and consumption of pork.

⑨ ___ Americans eat more mutton than lamb.

⑩ ___ Sheep can make better use of lower-quality forage than can cattle.

Multiple Choice

① The largest segment of the meat industry is
    a. beef
    b. poultry
    c. fish

② About 90 percent of all feed consumed by sheep is
    a. mutton
    b. corn
    c. roughage

3 The major grain fed to swine is
   a. corn
   b. wheat
   c. roughage

4 Horses that weigh 900 to 1,400 pounds are called
   a. light
   b. purebred
   c. crossbreed

5 The third largest spectator sport in the United States is
   a. horse racing
   b. football
   c. auto racing

6 Most rabbits are raised indoors in cages called
   a. hutches
   b. sires
   c. dams

7 A substance found in meat that can cause heart problems is called
   a. mutton
   b. roughage
   c. cholesterol

8 Over three-quarters of all of the money earned from the meat industry comes from
   a. chicken
   b. beef
   c. turkey

9 The mild winters of the South are ideal for
   a. calving
   b. racing
   c. wool production

10 The step between the weaning of the calves and the finishing or fattening of animals prior to slaughter is
   a. registration
   b. feedlot operations
   c. stocker operations

Discussion

1. List some uses for animals.

2. Why are health authorities expressing concern over the consumption of so much meat in the United States?

3. What vitamins do meat and animal products provide?

4. How do livestock producers choose the breed they grow?

5. What are the four major segments of the beef industry?

6. What is an advantage of producing lambs?

7. Why is it better to raise sheep in a drier climate?

8. How is wool classified?

9. Why did the number of horses and mules used in this country decline?

10. What are the advantages of raising rabbits?

CHAPTER 15

The Dairy Industry

Student Objectives

When you have finished studying this chapter, you should be able to:

- Explain the importance of the dairy industry.
- Discuss the controversy surrounding the use of BST.
- Tell how milk is produced.
- Explain how milk is processed and graded.
- Discuss the cheese making process.
- Tell how yogurt is made.

Key Terms

dairy products	fermentation	pasteurization
selective breeding	milkfat	cheese
BST	homogenization	yogurt

270

FIGURE 15–1
Around 13 percent of all sales of farm commodities are from dairy products. *Courtesy of James Strawser, The University of Georgia.*

A large portion of agriculture is the dairy industry. This industry includes all of the products made from milk. Around 13 percent of all sales of farm commodities are from **dairy products** (Figure 15–1). Americans use a lot of dairy products. On average, we consume 28.8 gallons of milk, 26.8 pounds of cheese, 16 pounds of ice cream, 4.3 pounds of butter, and 4.3 pounds of yogurt per person, per year. These products are high in protein and contain essential minerals and vitamins. Milk has been described as nature's most perfect food. It is the only substance intended by nature for no other purpose than as food. Milk is obtained from cows that produce it to feed their calves (Figure 15–2). Because it is the only food a very young animal receives, it must be filled with nutrients. Humans make use of this nutritious food.

Selective breeding of dairy cattle has developed cows that produce much more milk each year than a calf could consume. This allows us to use the surplus for our food. There are about half the number of dairy cows in this country than there were in 1950. However, the amount of milk produced remains about the same. This means that modern

FIGURE 15–2
Milk is intended by nature to be a food for young animals. *Courtesy of Getty Images.*

dairy animals produce twice the milk their ancestors did.

In recent years, consumers have been concerned over a substance given to dairy cattle to increase production. This substance, called bovine somatotropin or **BST**, is a naturally occurring hormone that helps to stimulate milk production. Scientists have known for many years that cows that get additional amounts of BST significantly increase their milk production. Until recently, this substance was scarce and very expensive. However, because of genetic engineering, the hormone can now be produced quickly and cheaply. The use of BST has brought about controversy over the safety of the substance. Some say that no studies have been conducted to determine the long-term effects of drinking milk from cows that have been given BST. They point to the fact that large doses can cause inflammation of the cows' udders and that this may be a sign that the hormone is not safe. Proponents point out that the cows are only given very small doses and that no ill effects have been discovered. Also, the National Institute of Health and the Food and Drug Administration have declared BST to be safe for both cows and humans who consume the milk. All milk contains a small amount of BST. BST is a naturally occurring substance. Milk from cows given BST has only a minute amount more BST than milk from cows that have had no artificial BST. All BST in the milk is digested and none gets into the bloodstream of humans.

Milk Production

Most of the dairy cattle in the United States are Holstein (Figure 15–3). This breed produces more milk than most other breeds. Some breeds, like the Jersey, produce milk with a higher fat content. This was once a desirable characteristic. Today, however,

FIGURE 15–3
Most dairy animals in the United States are of the Holstein breed.
Courtesy of James Strawser, The University of Georgia.

consumers want foods with less fat content. Most of the milk produced in the United States comes from Wisconsin, California, New York, Minnesota, and Pennsylvania. These leading states have the climate necessary for cows to do well and large cities where the milk is marketed. However, every state produces and processes some milk.

Most dairy animals are fed on pastures. Grass makes a relatively inexpensive feed for the cows. Some dairies now feed their cows in lots rather than pastures. These cows are fed on a feed called silage. This feed is made by chopping green corn (ears, stalks, and leaves) (Figure 15–4). The chopped corn goes through a process known as **fermentation,** which is a chemical change, that helps to preserve the feed.

In order for cows to produce milk they must have a calf each year. Remember that the function of milk is to feed the young animal. The cycle of birth stimulates the mammary system of cows to produce milk. Most of the cows are bred artificially. This offers several advantages. By using artificial insemination (AI), producers do not have to go to the trouble and expense to keep a bull. Also, problems with sexually transmitted diseases are eliminated. AI gives the advantage of being able to select semen from a wide variety of superior bulls at a relatively low price. Daughters from these bulls can be used for replacement heifers and the quality of the herd can be raised.

When the calf is born, the first milk that is produced is different. This milk, called colostrum, is loaded with the mother's antibodies (Figure 15–5). These help the young animal to ward off disease during the first few days of life. Also, colostrum contains substances that help to open up the calf's digestive system and get it to function properly. Although this milk is essential to the young animal, it is considered unfit for human consumption. Cows

FIGURE 15–4
Dairy cows are fed on silage, which is made from chopped corn.
*Courtesy of James Strawser,
The University of Georgia.*

FIGURE 15–5
The first milk a newborn calf gets is called colostrum. *Courtesy of James Strawser, The University of Georgia.*

that have just given birth are separated and the colostrum is not mixed with milk intended for the market.

The cows are milked in an area called a parlor. These areas are scientifically designed so the cows can be handled gently. They come into the parlor and enter a station where they obtain feed. The cow stands still while she is being milked. A milker enters the parlor and milks a small amount of milk by hand into a container called a strip cup. This gives the milker an opportunity to check for problems in the milk. Lumps or blood in the milk could indicate a disease called mastitis. This disease is caused by injury to the cow's udder. Milk from cows with mastitis is not used for human consumption. Milking and discarding the first two or three squirts of milk also helps to cleanse the teat.

The udder is then washed using a warm water solution and dried. The teat cups are then attached and the milking begins. These cups are lined with a soft material that is attached to a tube that gently pulls a vacuum on the cup to draw milk from the teat (Figure 15–6). The teat cup fits snugly on the cow's teat and pulsates by means of a vacuum on the lining of the cup. The milk is all removed in 3 to 6 minutes depending on the individual cow and the amount of milk she gives. Care is given that the cups are left on the proper amount of time. Too little time and the udder will not be milked out, and too much time and injury to the udder can result. The teats are then treated with a disinfectant and the cows are released. The teat cups are cleaned between cows to prevent the spread of disease.

The milk is drawn through lines and into a holding tank. In modern dairies the cooling of the milk begins in the lines. The lines are refrigerated so that the milk is cooled as it is transported to the tank. Here it is rapidly cooled to around 40 degrees F to prevent the multiplication of bacteria and to prevent the milk from spoiling. After all the cows have been milked, the lines, teat cups, and all of the equipment are thoroughly cleaned. About every other day, the milk is picked up by a tanker truck and hauled to the processing plant. At the plant, the milk is tested for the number of bacteria, drug residue, and any indications of infection. Milk with high levels of bacteria, drug residues, or signs of infection is discarded.

When the milk arrives at the processing plant it is thoroughly filtered to remove any foreign particles. The milk is allowed to sit so the cream may be removed from milk to be sold as lowfat milk. Cream is the part of the milk that contains fat. As consumers are becoming more conscious of the amount of fat in their diet, they want milk that is lower in **milkfat** than whole milk. In recent years, sales of lowfat and skim milk have increased sharply (Figure 15–7). Lowfat milk is milk that has had the percentage of milkfat lowered to between 0.5 and 2 percent. Skim or nonfat milk is a milk containing less than 0.5 percent milkfat. The milkfat that is removed from the milk is then used to make ice cream and other cream products.

Whole milk contains around 4 percent milkfat. The globules of fat are what makes the cream float

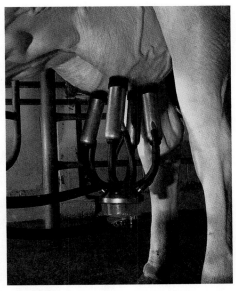

FIGURE 15–6
A milker gently attaches teat cups that draw out the milk. *Courtesy of James Strawser, The University of Georgia.*

to the top of raw, unprocessed milk. In a process called **homogenization**, the large cream globules are reduced in size to the size of the milk globules. The processed milk, called homogenized milk, will not separate out.

Career Development Events

The FFA offers many exciting and challenging experiences. Members who participate in team events spend many hours learning the basics in the classroom and practicing skills after school. Dairy Cattle Judging and Milk Quality and Dairy Food competition are two events that require cooperative efforts of team members. If you have an interest in dairy products, plan to own dairy cattle, have a Supervised Agricultural Experience Program (SAEP) in the dairy industry, or work on a dairy farm, you would be a good team member for these events.

The Dairy Cattle Judging Career Development Event trains students in the selection of a quality dairy herd. Practice and experience are the only ways to develop the skills needed to select quality animals. A well-prepared team should understand what a quality dairy animal looks like and be familiar with dairy herd improvement records. Most Dairy Judging teams consist of four FFA members. Usually, through a competitive process, the agriculture teacher selects a team to represent the FFA chapter in competition. The FFA Dairy Judging event requires each team member to know the breeds of dairy cattle and traits that yield high milk production. During the event, each member is responsible for evaluating and placing groups of animals in numerical order. The student must also give reasons for each animal's placement. After each team member has evaluated and placed all dairy groups, individual scores are calculated. The top three scores are added to determine the total team score and placement. All state winning teams advance to the National Dairy Judging Career Development Event.

Another FFA event that helps prepare students for careers in the dairy industry is the Milk Quality and Dairy Foods Event. Hundreds of products are manufactured by the dairy industry. Every dairy product is graded by its quality. It is important that workers in the dairy industry and consumers know what makes a high-quality product.

To participate in this event, a student must have a keen sense of sight, smell,

To kill any harmful organisms in the milk, the milk is first heated and then cooled in a process called **pasteurization** (Figure 15–8). In pasteurization, the temperature of the milk is raised and then is promptly cooled. This kills harmful bacteria that

and taste to distinguish grades of dairy products. Classroom study of milk content and flavor, classification of cheeses, and the parts and operation of milking equipment are essential. The skills required to rank dairy products in order of quality can only be developed by

practice. The Milk Quality and Dairy Foods Event is a great opportunity to prepare FFA members with skills required to be competitive dairy workers and knowledgeable consumers.

Students competing in the Milk Quality and Dairy Foods Career Development Event must distinguish between grades of dairy products. *Courtesy of Carol Duval, National FFA Organization.*

FIGURE 15–7
Milk is sold as whole, lowfat, and skim milk. *Courtesy of James Strawser, The University of Georgia.*

might cause illness in humans. The time and temperature must be precisely controlled to protect the nutritive value and flavor of the milk.

Milk is graded according to the dairy from which it came. Dairies that sell Grade A milk must pass rigid standards that include cleanliness and other conditions. Only Grade A milk can be used as fluid or beverage milk. Grade B milk can only be used for manufactured dairy products. As the production of Grade A milk far exceeds the demand for fluid milk, Grade A is used in processing also. For pricing

FIGURE 15–8
In the process of pasteurization, harmful bacteria are destroyed. *Courtesy of James Strawser, The University of Georgia.*

purposes, the milk is classified as Class I, II, or III. Class I is used for beverage consumption; Class II is used for manufacturing soft products such as ice cream, yogurt, and cottage cheese; Class III is used with Grade B milk in the processing of cheese, butter, and nonfat dry milk.

Cheese

The United States produces more **cheese** than any other country in the world. There are literally hundreds of types of cheese. They are broadly classified according to the type of milk used and the type of processing. Cheeses are often named for the town or region where they originated. Most of the cheese produced in the United States uses cow's milk. In some parts of the world, goat's milk (called Chevre cheese) or sheep's milk (called Roquefort) may be used in the process. Other cultures may make cheese from the milk of horses or camels.

Cheese contains one of the highest concentrations of nutrients of any food we eat. Only 3.5 ounces supplies about 36 percent of the protein, 80 percent of the calcium, and 34 percent of the fat recommended in our daily diet. It is rich in essential amino acids, minerals, and vitamins, and has a high caloric content.

Most cheese begins with milk that has been pasteurized to help prevent the multiplication of harmful bacteria. The milk is placed in a large vat and a bacteria culture is added (Figure 15–9). This culture is called a starter culture because it starts the process of fermentation, which changes sugars to acids. These acids cause the proteins in the milk to coagulate, that is, to form into a solid. To further the process, an enzyme called rennet is added. An enzyme is a substance that speeds up or stimulates a chemical process. Rennet is obtained from the stomachs of calves. During this step, large paddles

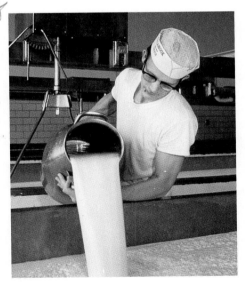

FIGURE 15–9
To begin the cheese making process, milk is placed in a large vat and starter cultures of bacteria are added. *Courtesy of Tillomook County Dairy.*

FIGURE 15–10
The solid is called curd and the liquid that is drained away is called whey. *Courtesy of Tillomook County Dairy.*

turn the milk to ensure that the starter bacteria and the rennet are evenly distributed.

The solid resulting from this step is called curd. The liquid is drained off and is called whey (Figure 15–10). Whey is used in processing human food and manufacturing animal feed. The curd is cut into small cubes to increase the surface area to allow the drainage of the whey. After the whey is drained off, the curd sits until it forms a solid mass. The curd is then heated to cause the curd to contract and further expel the whey. The amount of heat and length of time the cheese is heated depend on the type of cheese being made. The cheese is salted and pressed into a metal form or a cloth bag.

The final step in cheese making is the curing or ripening of the cheese. The cheese is placed in an environment that is controlled for temperature and humidity. The specific conditions vary with the type of cheese being made. During this time, enzymes produced from the starter bacteria change the flavor, texture, and appearance of the cheese. The cheese is packaged in a coating of paraffin or is wrapped in cloth or plastic. It is allowed to age a specific amount of time before it is sold (Figure 15–11).

Yogurt

Yogurt is also processed from milk. This food product has been around for hundreds of years. It has gained in popularity during the past few years. When frozen, it is delicious and is lower in calories than ice cream. Many people prefer frozen yogurt to ice cream. Yogurt comes in a wide variety of flavors and is often flavored with fruit.

Processing begins by heating concentrated milk. It is cooled rapidly and cultures of bacteria are added. These cultures cause the milk to ferment. In this process, the required acidity and the yogurt flavor are produced. The acidity helps to give the yogurt a tart flavor and also helps to preserve it. Before it is packaged, fruit or other flavorings may be added.

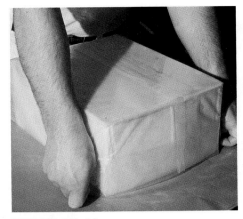

FIGURE 15–11
Cheese is coated, packaged, and stored for aging. *Courtesy of Tillomook County Dairy.*

Summary

Dairy is a very significant part of our agricultural industry. It provides many jobs in the production, processing, and distribution of milk and milk products. It also provides Americans with wholesome, nutritious, and good-tasting food. In the future, new uses may be found for milk products that will make the industry even more important.

15 CHAPTER REVIEW

Student Learning Activities

1. Visit a large grocery store and make a list of all the different types of dairy products that are available. Interview the manager to find out which sell the best.

2. Bring in samples of whole, lowfat, and skim milk. Blindfold class members and see if they can identify each by taste. Ask them how each tastes and record their responses.

3. Choose a particular type of cheese and write an essay on the cheese. Include how the cheese is different from other cheese, where and how it originated, where it is made, and the process used in the manufacture.

4. Collect all the information you can about the use of BST with dairy cattle. Decide whether or not you believe the product should be used. Defend your position to the class.

True/False

1. ___ Around 50 percent of all sales of farm commodities are from dairy products.

2. ___ Milk has been described as nature's most perfect food.

3. ___ There are about half the number of dairy cows in this country than there were in 1950.

4. ___ BST is a hormone that is very expensive to produce.

5. ___ All milk contains a small amount of BST.

6. ___ Only a few states process and produce milk.

7. ___ In order for cows to produce milk they must have a calf each year.

8. ___ Colostrum is not used for human consumption.

9 ___ The United States produces more cheese than any other country in the world.

10 ___ Milk is the only substance that is intended by nature for no other use than as food.

Multiple Choice

1 A very large portion of animal agriculture is the
   a. dairy industry
   b. pesticide industry
   c. wheat industry

2 The dairy industry includes all of the products made from
   a. cows
   b. colostrum
   c. milk

3 Dairy products are high in
   a. hormones
   b. mastitis
   c. protein

4 Most of the dairy cattle in the United States are
   a. Holstein
   b. Angus
   c. Sires

5 A relatively inexpensive feed for cows is
   a. wheat
   b. grass
   c. grain

6 Most of the cheese produced in the United States comes from
   a. goats
   b. cows
   c. artificial dairies

7 The enzyme rennet is obtained from calves'
   a. udders
   b. kidneys
   c. stomachs

8 Colostrum is loaded with
   a. chemicals
   b. antibodies
   c. acids

9 Mastitis is caused by injury to the cow's
   a. stomach
   b. udder
   c. kidneys

10 Whole milk contains about 4 percent
   a. colostrum
   b. calcium
   c. milkfat

Discussion

1 What are the advantages of artificial breeding?

2 Why is colostrum good for newborn calves?

3 How is milk classified?

4 What animals are used for their milk?

5 What nutrients are supplied by cheese?

6 What purpose does acidity have in yogurt?

7 Why is the dairy industry so important?

8 What does fermentation do to milk?

9 Why are bacteria added to milk to make yogurt?

10 Why was there concern about the hormone BST?

CHAPTER 16

The Poultry Industry

Student Objectives

When you have finished studying this chapter, you should be able to:

- Explain the importance of poultry to humans.
- List the different segments of the poultry industry.
- Explain how eggs are hatched commercially.
- Describe how broilers are produced.
- Explain how eggs are produced.
- Discuss the growing of turkeys and other poultry.

Key Terms

game birds	nutrition	brooder
broilers	incubator	layers

FIGURE 16–1
Poultry is lower in calories and cholesterol than beef or pork.
Courtesy of American Broiler Council.

Humans have grown poultry for almost as long as they have been involved with agriculture. The earliest archeological evidence reveals that poultry was a part of the human diet. All through the history of the world's great civilizations is evidence that people ate chickens and other birds.

Modern Americans also eat a lot of poultry. Each year, the average person in this country consumes almost 90 pounds of chicken and 22 dozen eggs. The amount of poultry meat consumed is rising each year. The reason is that poultry is considered to be healthier than some other meats. The amount of cholesterol and fat is lower in a pound of chicken or turkey than in a pound of red meat (Figure 16–1). The poultry industry includes the raising of chickens, turkeys, ducks, geese, **game birds** (birds hunted for food), and squab (pigeons). By far, the largest segment of the industry is the growing of chickens. This industry is divided into two parts—broiler production and egg production.

At one time most farms raised chickens for both meat and eggs. Females were kept to lay eggs and the males were butchered for meat. Today the industry is more specialized. Two different types of birds are produced. Through selective breeding, chickens have been developed that are thickly muscled and produce more meat. Layers have been developed that produce eggs daily (Figure 16–2). They are more slender than the birds raised for meat. Layers convert more of their energy into egg production than do meat type birds.

The Broiler Industry

Broilers are chickens that are produced for meat. They are generally marketed at about six to eight weeks of age and weigh 3 to 5 pounds. Since 1950, the production of broilers has increased dramatically. Now we can grow a broiler that is heavier on half the feed in half the time (Figure 16–3). This is mostly because of research in **nutrition** and management and also to selective breeding. This means a real savings for the consumer. The efficiency with which these animals can be grown means that they can be sold cheaper.

Most of the nation's broilers are grown in the southern states. The leading states in broiler production are Arkansas, Georgia, and Alabama. In these states the winters are mild and the chickens can be grown at a cheaper cost. Most broilers are raised on contract. This means that the producer supplies the house, utilities, and labor. A company supplies the producer with chicks, feed, medication, and other supplies. The company pays the producer for the broilers at the end of the grow-out period. This substantially reduces the risk for the producer.

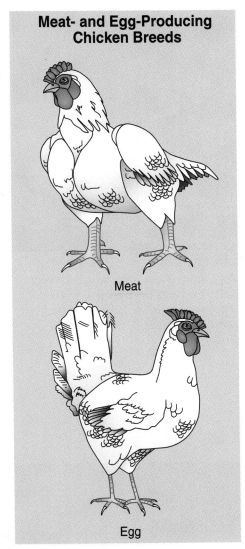

Meat- and Egg-Producing Chicken Breeds

Meat

Egg

FIGURE 16–2
Broiler chickens are a different type than layer chickens.

Responsibilities of The producer & the company when broilers are raised on Contract. #15

#14 Why most broilers are white

FIGURE 16–3
We can grow a heavier broiler in half the time on half the feed than we could in 1950. *Courtesy of James Strawser, The University of Georgia.*

Hatching the Chicks

The first step in the production of broilers is the hatching of chicks. Breeder birds are carefully selected for rapid growth, good feed efficiency, and heavy muscling. Almost all of the broilers produced are white (Figure 16–4). This is because birds that are dark in color have spots of color or pigmentation left where the feathers are removed after slaughter. These spots do not lower the quality of the meat in any way, but consumers are reluctant to buy chicken with spots on the skin.

Eggs are hatched in large commercial hatcheries. Eggs that are to be hatched have to be fertile. This means that the hens have to be bred by roosters or they are artificially inseminated. Hens that lay eggs for hatching are housed in facilities with laying boxes (Figure 16–5). These boxes are designed to keep the eggs clean. It is essential that hatching eggs be clean. Even a small speck of foreign material on the shell can contain millions of

FIGURE 16–4
Almost all of the broilers produced are white. *Courtesy of James Strawser, The University of Georgia.*

FIGURE 16–5
Hens that lay eggs for hatching are housed in facilities with laying boxes. *Courtesy of James Strawser, The University of Georgia.*

microorganisms that can cause problems. Eggs that are dirty are not used for hatching. Dirt or any foreign material that must be scrubbed from the eggs usually renders the eggs unfit for hatching. The washing or scrubbing of the eggs removes the protective coating from the eggs and presses the dirt into the pores of the eggs. Hatching eggs are not allowed to become wet. Moisture on the eggs can also encourage the growth of bacteria (Figure 16–6).

Before they leave the farm where they are produced, hatching eggs are sorted to remove dirty, undersized, oversized, misshapen, cracked, or defective eggs. They are then fumigated to kill harmful organisms on the surface of the eggs. This process is precisely regulated to prevent harm to the eggs. The eggs are never allowed to become chilled and are stored at 70 to 80 degrees F until they are placed in the hatchery.

The eggs are transported to the hatchery on special racks that protect the eggs. At the hatchery, the eggs are removed from the carts and placed in the **incubator**. Here the temperature is gradually increased to prevent the eggs from sweating. Sweating is the condensing of water vapor on the surface of the eggs and occurs when the temperature of the eggs is raised too rapidly. If moisture collects on the surface of the egg, bacteria may begin to grow and thrive.

In commercial hatcheries, the eggs are incubated in two separate rooms, the setting room and the hatching room. The eggs are placed in the setting room incubator and closely monitored for temperature and relative humidity (Figure 16–7). The eggs are turned every day to ensure that a high percentage of the eggs hatch. The eggs remain in the setting room incubator until one to two days before they are ready to hatch. They are then placed in the hatching room. The temperature is

FIGURE 16–6
Hatching eggs must be kept clean and dry. *Courtesy of James Strawser, The University of Georgia.*

FIGURE 16–7
Eggs are placed in an incubator and are carefully monitored for temperature and relative humidity. *Courtesy of James Strawser, The University of Georgia.*

lowered slightly and the eggs hatch in chick holding trays.

Growing the Broilers

After the chicks are hatched they are cleaned and dried. They are then placed in a warm environment. When they are about a day old, they are vaccinated and debeaked. In debeaking, the tip of the beak is removed. This helps to prevent the chicks from pecking each other (Figure 16–8). It does not harm the chicks and prevents problems throughout the life of the chickens.

The chicks are placed in cardboard or plastic ventilated boxes and shipped to the producers. The chicks are placed in houses that hold from 10,000 to 50,000 or more chickens depending on the size of the house. The houses are well insulated and well lit. Lights are kept on around the clock. This helps prevent a problem called cannibalism. For some reason, the chickens may begin to peck certain birds. This causes injury and often death. By debeaking the chicks and leaving the light on, cannibalism can be lowered. The floor is covered with

FIGURE 16–8
Chicks are debeaked to prevent them from pecking each other. *Courtesy of James Strawser, The University of Georgia.*

FIGURE 16–9
Chicks are kept under a brooder to keep them warm. *Courtesy of James Strawser, The University of Georgia.*

an absorbent litter consisting of shavings, sawdust, or other material.

At the start the baby chicks are placed under a **brooder**. Brooders are large pan-shaped heaters used to keep the chicks warm during the first days in the house (Figure 16–9). Brooders are usually suspended from the ceiling and can be raised or lowered depending on the temperature and the size of the chicks.

Water is supplied by suspended waterers that the chicks quickly learn to use. They peck the nipple on the bottom of the waterer and obtain as much water as they need. Feed is given to the baby chicks by hand when they are small but is fed through automatic feeders as the birds get older (Figure 16–10). The feed is brought to the birds by means of a conveyer chain in the bottom of the trough. Every day, the equipment and birds are checked to ensure that the equipment is functioning properly and that the birds are doing well.

The birds are generally kept in the broiler house from seven to eight weeks. At this time, they weigh around 4.5 pounds and are ready for market. They are usually caught at night when the birds are less

FIGURE 16–10
The chickens are fed through automatic feeders and waterers. *Courtesy of James Strawser, The University of Georgia.*

active. They are put into coops and loaded on a truck for transportation to the processing plant. The producer then begins to clean the house to get ready for the next batch of chicks. <u>Because</u>

Career Development Events

The FFA offers agriculture students opportunities to apply classroom lessons and laboratory skills through competition. Matthew joined the FFA his first year in high school. One of his classes in agriculture was poultry science. Matthew's experience and classroom knowledge prepared him to be a winner in several FFA competitions. Matthew tried out for the Poultry Judging career development event. FFA Poultry Judging is a team event. Matthew made the team along with three other FFA members. The team members spent many hours preparing for the area competition. After school, they practiced judging and placing classes of live birds, dressed poultry, interior and exterior of eggs, and processed chicken parts. At home they each studied assigned topics in the poultry science manual that their agriculture teacher would drill them on the following day. Matthew practiced public speaking skills. After weeks of practice and study, the team competed in the area FFA Poultry Judging event. Each of the four members had to judge and give

reasons for classes of production hens, pullets, ready-to-cook broilers, chicken patties, interior eggs, and exterior eggs. They also had to identify ten broiler parts and place four cartons of eggs with an explanation of how they placed them. At the end of the event, each FFA member had to answer a 30-question test relating to management practices used in the poultry industry. After all teams completed each section of the event, individual scores were calculated. The top three individual scores for each team were added to determine the team score and placement. Matthew's team placed second and has the opportunity to advance to the state event. Matthew's interest in the processing and marketing of poultry, development of a supervised poultry project, and devotion of time and energy has made him a winner and prepared him for a future career.

The Poultry Proficiency is one of many agricultural proficiency awards presented at the FFA's annual banquet. Applicants applying for the poultry production proficiency award must be

the manure in the litter has a very high concentration of nitrogen and other elements necessary for plant growth, the litter is a valuable source of fertilizer.

involved in activities and enterprises requiring the application of classroom lessons to the profitable management of a poultry operation. All FFA members who had a Supervised Agricultural Experience Program (SAEP) in the area of poultry competed for this award.

Maintenance of a complete set of records including income and expenses, inventory of poultry and related equipment, and knowledge of the poultry industry help give students the winning edge.

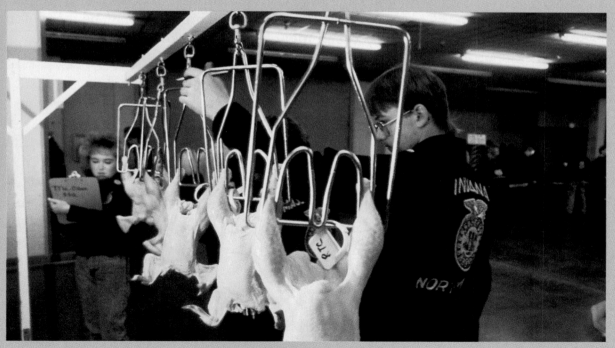

In the Poultry Judging Career Development Event, FFA members must determine the quality of various poultry products. *Courtesy of Carol Duval, National FFA Organization.*

Processing the Broilers

At the processing plant, the birds are inspected while they are alive. If they show any signs of ill health they are condemned. This means that they cannot be used for human consumption.

After inspection, the chickens are slaughtered and the feathers are removed. The intestines and other internal organs are removed. The carcasses are then washed and inspected. Both the carcass and the internal organs are inspected for signs of disease or any other problems (Figure 16–11). If any problems are found, the carcass is condemned and the meat is not used for human consumption.

After the birds are dressed, they are prepared for sale. They may be packaged whole or they may be cut up into several pieces. The trend is toward selling different parts of the carcass. For example, a package may contain only breasts, legs and thighs, or wings. Some of the broilers may be used for processed meat such as chicken franks, sausage, or

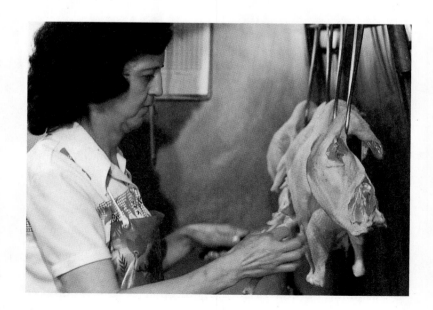

FIGURE 16–11
Carcasses are inspected for any signs of disease or other problems. *Courtesy of James Strawser, The University of Georgia.*

bologna. Others may end up as chicken pot pies or as prepared frozen dinners. ~~Stop~~

The Layer Industry

The other large segment of the poultry industry is the production of eggs. As mentioned earlier, hens or **layers** have been developed through selective breeding that produce many more eggs than their ancestors. Today, the average layer produces over 250 eggs per year. Some layers produce brown eggs and some produce white eggs, but the vast majority of the eggs sold in the United States are white. Consumers simply prefer to buy white eggs. Most of the layers in this country are housed in cages within a large facility. The cages hold from 2 to 12 birds.

The hens are fed by means of an automated conveyer that carries the feed directly in front of the hens (Figure 16–12). Water is supplied by means of a narrow free-running trough or a nipple waterer that the hens learn to peck to obtain water. As the eggs are laid, they roll onto a conveyer that periodically moves the eggs to a collection point where a worker gathers them and places them in flats (Figure 16–13). Dirty eggs are separated out and the eggs are refrigerated.

At the processing and packing plant, the eggs are coated with a thin coat of light mineral oil. This coating prevents carbon dioxide from escaping from within the egg. This helps prevent the eggs from spoiling before they are consumed. The eggs are checked for cracks and interior spots in a process called candling (Figure 16–14). The eggs are passed over an intense light in a dark room and any blood spots or cracks in the shell will show up in the light. The eggs are then graded according to shape and size. Eggs of similar size are packaged together.

FIGURE 16–12
Layers are housed in cages where they are fed and watered by automation. *Courtesy of James Strawser, The University of Georgia.*

FIGURE 16–13
Eggs are brought from the laying cages on a conveyer. Sometimes they are gathered and packaged at the farm. *Courtesy of James Strawser, The University of Georgia.*

FIGURE 16–14
In a process called candling the eggs are checked for blood spots or cracks. *Courtesy of James Strawser, The University of Georgia.*

The eggs are packaged and sent to the retail market or are sent to a processing plant where they are broken and processed.

Turkey Production

Turkey production is on the increase in this country. This is because of the many new ways of using turkey meat. Luncheon meat, sausages, wieners, and products resembling ham are made from turkey. Turkey meat is less expensive, and has fewer calories and cholesterol than many other meats. It also tastes good and is nutritious.

Turkeys are the descendants of the wild turkeys that were native to the Americas. Modern turkeys are very different from the wild variety. Like chickens, most domestic turkeys are white. Consumers do not want the coloring in the skin that comes with the bronze-colored turkeys. Also turkeys raised for meat are much more muscular. The modern broad-breasted turkey is much more efficient at producing meat than its wild ancestors.

The majority of the turkeys in the United States are grown in the western part of the North Central region, the South Atlantic region, and the Pacific region. Unlike chickens, turkeys seem better able to tolerate cold weather than hot weather. Most are produced by small operations of 30,000 birds or fewer.

There are two major ways of growing turkeys—in confinement houses and on ranges (Figure 16–15). Confinement houses offer the advantages of environmental control of temperature and humidity. The open range offers the advantage of being less expensive. Turkeys on range can stay outdoors completely or they are provided with housing where the birds can get shelter when they want. Producers usually move the range every three to four years to help prevent problems with disease and parasites.

Other Poultry

Although chickens and turkeys make up by far the largest portion of the poultry industry, other types of poultry are also grown. In some parts of the world, poultry such as ducks and geese make up a major portion of the total poultry output. In China and Southeast Asia, ducks and geese make up a large part of the overall diet (Figure 16–16).

FIGURE 16–15
Turkeys may be grown on the range or in a house. *Courtesy of James Strawser, The University of Georgia.*

FIGURE 16–16
In some parts of the world, ducks and geese make up a major part of the overall diet. *Courtesy of Robert Newcomb, The University of Georgia.*

FIGURE 16–17
Pheasants and quail are raised for use in restaurants. *Courtesy of International Quail, Greensboro, GA.*

In some ways, these birds are actually easier to raise than chickens and turkeys. They withstand harsh weather better and resist disease better than turkeys and chickens. Another advantage of growing geese and ducks is that the feathers are used to make bedding and other goods. In this country, most ducks and geese are raised in small flocks by hobbyists or part-time producers. Most of the meat sold goes to the restaurant trade or gourmet food market.

About the only other poultry production of any significance is that of growing quail and pheasant. Both of these birds are grown for the restaurant trade and the gourmet food market (Figure 16–17). Also, they are raised for restocking wildlife areas. Each year thousands of quail and pheasants are released in the wild to provide birds for hunting. This helps replenish the areas with game that is difficult to produce in the wild.

Summary

As the world's population increases, the growing of poultry will become more important. Poultry is efficient to raise and most cultures eat poultry. Many developing countries will depend on poultry to help in the development of their agriculture. In America, we will continue to eat more poultry and poultry products.

CHAPTER REVIEW

Student Learning Activities

1. Make an inventory of all the foods in your house that contain eggs. Do not forget foods such as pastries and mayonnaise. Share your list with the class.

2. Over a period of a week, keep track of all the meat products eaten by your family. Calculate the percentage of meals that contain poultry products. Include processed meats such as luncheon meats and hot dogs that may contain chicken or turkey.

3. Research the possibilities of raising exotic birds such as emus and ostriches. Provide a rationale for growing the birds and the advantages and disadvantages. Report to the class.

4. Go to the Internet and find the latest information on eating eggs in the diet. What conclusions can you draw?

True/False

1. ___ Humans have been growing poultry for almost as long as they have been involved with agriculture.

2 ___ Each year, the average person in the United States consumes almost 90 pounds of chicken and 22 dozen eggs.

3 ___ At one time, most farmers raised chickens for both meat and eggs.

4 ___ Most of the nation's broilers are grown in the northern states.

5 ___ Almost all of the broilers produced are tan.

6 ___ The vast majority of the eggs sold in the United States are white eggs.

7 ___ Turkeys cannot tolerate cold weather as well as chickens.

8 ___ In China and Southeast Asia, ducks and geese make up a large part of the overall diet.

9 ___ Chickens make up the largest part of the poultry industry. The second most widely produced birds are geese.

10 ___ Each year thousands of quail and pheasants are released into the wild to provide birds for hunting.

Multiple Choice

1 The meat that is considered to be healthier than most meats is
   a. beef
   b. pork
   c. poultry

2 The largest segment of the poultry industry is raising
   a. turkeys
   b. fish
   c. chickens

3 Raising chickens is divided into two parts, broiler production and
   a. egg production
   b. candling
   c. debeaking

4 Chickens that produce eggs daily are called
   a. layers
   b. brooders
   c. hatchers

5 Chickens produced for meat are called
   a. layers
   b. hatchers
   c. broilers

6 One of the first steps in the management of broilers is
   a. candling
   b. debeaking
   c. hatching

7 Eggs that are to be hatched must be
   a. fertile
   b. debeaked
   c. fumigated

8 At processing and packaging, eggs are lightly coated with
   a. cooking oil
   b. olive oil
   c. mineral oil

9 Candling is a process that checks for defects in
   a. feather pillows
   b. eggs
   c. fish

10 Areas designed to keep eggs clean are
   a. laying boxes
   b. incubators
   c. conveyors

Discussion

1 How are breeder birds selected?

2 Why are almost all of the broilers white?

3 Why is it essential that hatching eggs be clean?

4 What are the two rooms in a commercial hatchery?

5 Why are lights kept on in a chicken house?

6 Why is the litter in a chicken house a valuable fertilizer?

7 Why are eggs coated with mineral oil?

8 What are the two major ways of growing turkeys?

9 What animals are raised in the poultry industry?

10 What is meant by raising broilers on contract?

CHAPTER 17

The Science of Aquaculture

Student Objectives

When you have finished studying this chapter, you should be able to:

- Discuss why aquaculture is a growing industry.
- Explain why fish are efficient agricultural animals.
- Explain the steps involved in the production of both catfish and cool-water fish.
- Discuss the production of crustaceans for human consumption.
- Describe the nonfood segments of aquaculture.

Key Terms

aquaculture	roe	cool-water fish
ectothermic	fry	crustaceans
warm-water fish	fingerlings	ornamental fish
brood ponds	dissolved oxygen	

FIGURE 17–1
The amount of fish and seafood eaten by Americans is increasing each year. *Courtesy of John Wazniack, Louisiana State University Agricultural Center.*

One of the newest and most rapidly developing parts of agriculture is that of growing animals in water. This science is called **aquaculture**. Aqua means water and culture means growing or caring for. Many of the foods we eat were grown in the water. You probably ate some type of fish this past week. Perhaps you enjoy seafood such as oysters or shrimp. The amount of fish and seafood eaten by Americans is increasing each year (Figure 17–1). Experts tell us that eating fish is nutritious and makes a good addition to the diet. It is high in protein and low in calories. The fat content is lower than that of red meat, and this helps prevent health problems. As people become more health conscious, the demand for fish and seafood will increase.

People have grown fish for food for thousands of years. The ancient Chinese raised fish, as did the Egyptians and later the Romans. This provided these people with a ready supply of fish. Those of you who enjoy fishing know that on certain days, even the best angler cannot catch any fish. If the fish have been raised in small ponds, they can be caught at any time.

FIGURE 17–2
Up until a few years ago, most of the fish we ate came from the ocean. *Courtesy of James Strawser, The University of Georgia.*

Only during the past 30 years has aquaculture become a major industry in this country. Up until that time, most of the fish sold in the markets were caught from the wild (Figure 17–2). Large seagoing trawlers harvested the fish from the oceans. This supplied the world with a relatively inexpensive source of fish. However, as the world population increased and the demand for fish increased, more and more fish were caught from the ocean. A few years ago, the maximum amount of fish that could be taken from the oceans in a year was reached. The oceans simply cannot sustain enough fish to meet the demand. This created a greater demand for fish than there was a supply of fish. As a result the price of fish went up. To provide people with a supply of fish, fish farms came into production. # 2

At first, not much was known about growing fish for food. But scientists at several of our universities conducted extensive research in how to produce fish. This resulted in an industry that is now providing us with an inexpensive supply of fish.

Fish are very efficient farm animals. This means that a pound of weight can be put on these animals with less feed than other agricultural animals. Fish

#3

FIGURE 17–3
Aquatic animals that are suspended in water use less energy moving about. *Courtesy of James Strawser, The University of Georgia.*

can gain a pound of body weight on less than two pounds of feed. A steer may require 8 or 9 pounds of feed to gain a pound of body weight. This is partly because fish are **ectothermic**. This means that they depend on their environment for their body temperature. In other words, the water gives them their body temperature. Most other agricultural animals, such as chickens, pigs, and cattle, are endothermic and rely on heat generated from their feed to maintain their body temperature. Land animals also have to use a lot of energy moving about. Aquatic animals are suspended in the water and use less energy in moving about (Figure 17–3). All this saved energy can be put into growing and gaining weight.

Catfish Production

By far the most important food fish grown in this country is the catfish. In 1960, there were only 600 acres of commercial catfish ponds in the United States. Today, there are over 150,000 acres of catfish ponds. These fish are raised in the southern part of the country because they are **warm-water fish**.

FIGURE 17–4 #5
Catfish has replaced cotton as the number one crop of the Mississippi delta region. *Courtesy of James Strawser, The University of Georgia.*

FIGURE 17–5
Catfish can tolerate relatively low levels of oxygen. *Courtesy of Evelyn Barnes, The University of Georgia.*

In fact, catfish grow best when the water temperature is around 85 degrees F. Mississippi leads the nation in catfish production. For many years, the fertile Mississippi delta region produced cotton as the major crop. Today, catfish has replaced cotton as the number one cash-producing agricultural enterprise (Figure 17–4).

Catfish are different from most other fish. They have a smooth skin instead of scales. They also have long whiskers (known as barbels) extending from their mouths. They are meaty and have a mild flavor. They can be cooked in a variety of ways ranging from deep fat frying to broiling. Also, catfish fit well into farming operations because they are hardy and tolerate relatively low oxygen content in the water (Figure 17–5).

The production of catfish begins with the breeding process. Through a system called selective breeding, superior fish are used to provide eggs for

FIGURE 17–6
The female catfish lays a mass of eggs called roe.

hatching. Factors such as growth rate and meatiness of the fish are used to determine which fish are superior. Although catfish have been known to reach a weight of over 100 pounds, the best size for a brood fish seems to be around 3 to 5 pounds.

The fish are placed in ponds less than 6 feet deep called **brood ponds**. These ponds are used to keep fish that breed and lay eggs. Boxes are supplied for the fish to use as a nest for spawning and are submerged. The female lays several thousand eggs (called **roe**) at a time (Figure 17–6). After the female lays a mass of roe in the nesting box, the male fertilizes the eggs by depositing sperm into the mass of eggs. In the wild, the male cares for the nest until the eggs hatch.

In aquaculture operations, the producer takes the eggs from the nest soon after they are laid and fertilized. The eggs are placed in a hatching tank where paddles rotate and gently move the water. After the eggs hatch, the small fish, called **fry**, are placed in long tanks. In these tanks the fry are fed and cared for until they become **fingerlings**. Fingerlings are small fish that are from 2 to 6 inches in length. This size fish is what is used to stock the ponds (Figure 17–7).

The grow-out ponds are about 4 to 6 feet deep and are usually around 1 to 2 acres in area. The smaller ponds make harvest easier. In these ponds the fish are fed on a protein-rich diet. The feed floats on the water so the fish have to come to the surface to get it. This allows the producer to see the fish each day as they are fed. An alternative is to raise the fish in cages that are submerged in the water. This gives the advantage of having the fish together in one place. This allows easy feeding, care, and harvest (Figure 17–8). Also the cage offers protection from predators.

During the period when the fish are growing, the producer keeps careful check on the pond. Fish, like

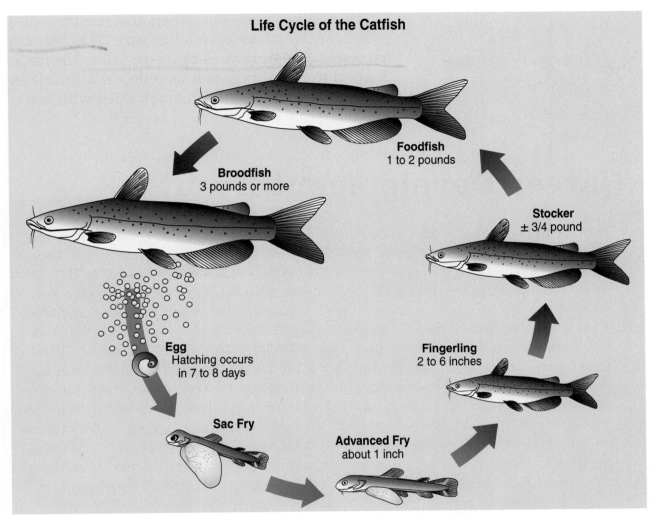

Life Cycle of the Catfish

Foodfish
1 to 2 pounds

Broodfish
3 pounds or more

Stocker
± 3/4 pound

Egg
Hatching occurs
in 7 to 8 days

Fingerling
2 to 6 inches

Sac Fry

Advanced Fry
about 1 inch

FIGURE 17–7
The life cycle of the catfish.

all other animals, must have oxygen. Their oxygen comes from the oxygen that is dissolved in the water. Water can run out of **dissolved oxygen**, and if this happens, the fish will die. The producer checks the oxygen level of the water to ensure that the fish have enough. Loss of oxygen occurs most often during hot, dark nights when plants in the water are not giving off oxygen. The producer can put oxygen back into the water by using mechanical aerators. These devices sling water into the air where oxygen becomes mixed with the water (Figure 17–9).

The fish are harvested when they weigh 1 to 2 pounds. Consumers prefer this size. The fish are harvested using a large net called a seine. The net is dragged through the water to collect the fish. They are then placed in a tanker truck filled with water

Career Development Events

The Speciality Animal Production proficiency is one of the many awards available to FFA members. This award is presented to individual FFA members who have effectively produced and marketed such animals as rabbits, fish, dogs, ducks, and bees. The student must have a Supervised Agricultural Experience Project (SAEP) in the selected special animal area. The breeding, feeding, and marketing of fish is an excellent project.

Pam's agriculture teacher invited her to join the FFA when she was in the ninth grade. Pam earned the Greenhand degree and attended FFA camp that first year. Her sophomore year, Pam wanted to be a part of the Supervised Agricultural Experience Program (SAEP). After consulting with her parents and agriculture teacher, she decided to grow tropical fish in tanks she kept in her basement. Pam signed up for an aquaculture class and learned how to select, feed, and adjust the environment to produce healthy stock. She purchased the necessary equipment to

perform water quality tests and determine weight gains of her fish. In the second year, Pam produced small fish that were sold to local pet stores. Pam's good management practices produced attractive, healthy fish. With the assistance of her agriculture teacher, Pam continued her aquaculture project for three years. Keeping an inventory of equipment, feed, supplies, and the fish was a constant job. Pam used a computer to track the amount of money she spent and made to calculate each year's net income. At school, Pam won first place in the science fair with a water quality project that she tested in controlled tanks with some of her fish. Pam sang in the school chorus and played soccer for the YMCA. By her senior year, Pam had earned her Chapter and State degrees, served as the chapter reporter, and participated in several FFA events. Becoming involved in chapter activities, developing an outstanding supervised experience program, keeping accurate records and inventories, showing the

containing a small amount of salt. The salt has a calming effect on the fish. It is important that the fish reach the processing plant alive (Figure 17–10). Once fish die, they spoil very rapidly. Packers clean and process the fish as quickly as possible. This

practice of required skills, and preparing a neat and complete application are required for winning proficiency awards. Pam won her chapter proficiency award in Specialty Animal Production and received a medal at the annual FFA banquet. Students who win chapter awards can advance to the district and state competitions. The National FFA Foundation awards plaques and prize money for state and national proficiency winners.

Students with aquaculture projects may qualify for the Specialty Animal Production Proficiency Award. *Courtesy of Blane Marable, Morgan Co. High School, Madison, GA.*

FIGURE 17–8
Raising fish in cages makes them easier to feed and harvest.
Courtesy of James Strawser, The University of Georgia.

ensures that consumers receive a fresh, high-quality product.

Cool-Water Fish

In the cooler northern portion of our country, trout, Atlantic salmon, and striped bass are raised. These are known as **cool-water fish** because they thrive in water that is below 65 degrees F. These fish require free-flowing clean water. They must also have a higher oxygen content in the water than catfish. For these reasons, trout and salmon are usually raised in concrete structures called raceways. Here water flows constantly (Figure 17–11). This not only helps keep the water clean but helps keep the oxygen levels high.

The spawning of cool-water fish is different than that of catfish. The procedure usually is aided by a technician. When the female is ready to lay eggs, the technician gently applies pressure to the lower abdomen of the fish. This causes the eggs to ooze out into a container. A similar process is then applied to the male fish, and sperm is collected. The sperm is then applied to the eggs for fertilization.

FIGURE 17–9
Mechanical aerators sling water into the air to add oxygen.
Courtesy of James Strawser, The University of Georgia.

The eggs are hatched in jars in which the water is controlled for temperature and oxygen level. When the fry hatch, they are raised in tanks until they are large enough to stock in the raceways.

Growing Crustaceans

Crustaceans are aquatic animals that have an exterior skeleton (exoskeleton) and three body parts—the head, thorax, and abdomen. Included are shrimp, crayfish, and prawns. These animals have been used as food for thousands of years. Although tons of these animals are harvested in the wild each year, many are raised by producers.

Crayfish are raised in several states. The leading states are Louisiana and Texas. Most are raised in rice fields. The rice fields have to be flooded for a length of time, and during this time the crayfish are raised in the water (Figure 17–12). They eat the decaying plant material left from growing the rice. They also eat worms and insect larvae. Crayfish are harvested using baited traps that are placed in the water. They are then placed in wet bags that are

FIGURE 17–10
Fish are delivered to the market alive. *Courtesy of James Strawser, The University of Georgia.*

FIGURE 17–11
Cool-water fish are raised in raceways of moving water. *Courtesy of The University of Georgia.*

FIGURE 17–12
Crayfish are raised in flooded rice fields. *Courtesy of John Wazniack, Louisiana State University Agricultural Center.*

FIGURE 17–13
Shrimp are raised in aquaculture operations also. *Courtesy of John Wazniack, Louisiana State University Agricultural Center.*

kept moist and cool and sent to the processing plant. Many of the crayfish are sold live.

Worldwide there are over 35,000 shrimp farms. The major shrimp producing countries are located in the eastern hemisphere (Figure 17–13). China, Thailand, the Philippines, Taiwan, and Japan are major producers. Only about 20 percent of the world's supply is produced in the western hemisphere. Central and South America are the leading regions.

Shrimp farms are generally located near the ocean because shrimp have to be raised in salt water. The producers usually make use of the rising tide or flood gates to fill impoundments with salt water. Many of the same technologies used with fish, such as mechanical aeration, are used in shrimp production. The shrimp are harvested by using seines. They are then put on ice and shipped throughout the world.

Prawns are sometimes referred to as freshwater shrimp. These animals have to be grown in very warm water (75 to 95 degrees F). The young (called larvae) have to be grown in brackish water. Brackish means that the water has a salt content lower than ocean water and higher than fresh water. At the proper stage, they are placed in fresh water. Here they are raised until harvest.

Ornamental Fish

One of the fastest growing segments of aquaculture is the growing of **ornamental fish** or fancy fish (Figure 17–14). A favorite hobby of many people is raising tropical fish in home aquariums. These fish come in a wide variety of sizes, shapes, and colors. These fish can be very valuable. Because of their value, extra care and expense can be invested in their production. Most tropical fish are raised in what is called a closed-loop system. This system

FIGURE 17–14
The raising of ornamental fish is a growing industry. *Courtesy of James Strawser, The University of Georgia.*

consists of a large tank, a filtration system, and an aeration system. The fish must be kept at a constant temperature and the water must be clean. The spawning process is carefully monitored because many tropical fish eat their young. Once the young are hatched, they are grown to a certain size and shipped to pet stores all around the world.

An ornamental fish called the koi is gaining in popularity. This fish is a species of carp and has varying patterns of gold, white, and black colors. They are used in landscape ponds to decorate yards. These fish can be quite expensive. Breeding fish of the right color and color pattern have been known to sell for millions of dollars.

Summary

Aquaculture is a departure from the traditional concept of agriculture. This relatively new industry will likely grow even more as we enter the twenty-first century. Someday, aquaculture crops such as lobsters, bullfrogs, and other animals now harvested from the wild may be widely produced. Agricultural research will find a way to make this happen.

CHAPTER REVIEW

Student Learning Activities

1. Visit the local pet shop and interview the manager. Determine where the shop gets the fish it sells. Ask how they are ordered, shipped, and received. Also, find out how the fish are cared for. Report your findings to the class.

2. Visit the grocery store and compare the price of farm raised fish with fish caught from the wild. Also compare the cost to beef, pork, and poultry.

3. Go to the library and research a particular type of fish. Determine if the fish might have potential for commercial production. Write your findings in a report.

4. Locate information on how you can grow fish in tanks or aquariums in your classroom. Make a list of all the equipment you will need. What problems are encountered with growing fish in your classroom?

True/False

1. ___ One of the newest and most rapidly developing parts of agriculture is that of growing animals in water.

2. ___ Fish are not very efficient farm animals.

3. ___ Aquatic animals use less energy than land animals.

4. ___ The most important food fish grown in this country is the catfish.

5. ___ Catfish has replaced cotton as the number one cash crop in the Mississippi delta.

6. ___ Cool-water fish require a low oxygen content in water.

7. ___ Shrimp must be raised in salt water.

8. ___ One of the fastest growing segments of aquaculture is the growing of ornamental fish.

9 ___ Fish is high in protein and low in calories.

10 ___ The fat content of fish is lower than that of red meat.

Multiple Choice

1 The state that leads the nation in catfish production is
    a. Missouri
    b. Mississippi
    c. Florida

2 The production of catfish begins with this process:
    a. breeding
    b. stocking
    c. spawning

3 Small fish that are from 2 to 6 inches in length are called
    a. crustaceans
    b. fingerlings
    c. prawns

4 Most tropical fish are raised in what is called a
    a. closed-loop system
    b. brood pond
    c. raceway

5 Most crayfish are raised in _____ fields.
    a. corn
    b. wheat
    c. rice

6 Catfish eggs are called
    a. roe
    b. seine
    c. prawns

7 Cool-water fish thrive in water that is below
    a. 45 degrees F
    b. 65 degrees C
    c. 65 degrees F

⑧ Trout and salmon are usually raised in concrete structures called
    a. prawns
    b. raceways
    c. closed-loop systems

⑨ Crustaceans are aquatic animals that have
    a. endoskeletons
    b. epidermis
    c. exoskeletons

⑩ The best size for a brood fish is around
    a. 3 to 5 pounds
    b. 7 to 8 pounds
    c. 4 to 6 pounds

Discussion

① Why is aquaculture such a growing industry?

② Why did fish farms come into existence?

③ Why are fish considered efficient farm animals?

④ Why are catfish so popular?

⑤ What are raceways?

⑥ What are crustaceans?

⑦ What are the major shrimp-producing countries?

⑧ What is brackish water?

⑨ When does the loss of oxygen occur most?

⑩ How can oxygen be put back into water?

CHAPTER 18

Companion Animals*

Student Objectives

When you have finished studying this chapter, you should be able to:

- Discuss some of the ways companion animals are used to assist humans.
- Explain the benefits of hippotherapy.
- Define zoonoses and list several types.
- Identify several aspects of responsible companion animal ownership.
- List several careers that deal with companion animals.

Key Terms

companion animals	hippotherapy	parasite
service animals	zoonoses	

*This chapter was contributed by Jean Kilnoski.

Have you ever considered **companion animals** to be a part of agriculture? We would all agree that people who own pets are involved in raising the animals. Companion animals play an important role in the agricultural industry. People have always had a relationship with animals. In the last 50 years, our relationship with many domestic animals has changed. As our society has become more urban, the importance of these animals is measured less by the work they perform for us. Animals still have an important place in the lives of many people, however. They are more than pets; they are friends and companions. It is this special relationship between the animal and its owner that identifies a companion animal.

Companion animals are an important part of our society. In fact, pet ownership is at an all-time high. Almost 60 percent of American households have at least one companion animal. Virtually any kind of pet can be considered a companion animal. Dogs and cats are the most common, but companion animals may also be horses, rabbits, hamsters, gerbils, guinea pigs, and ferrets (Figure 18–1). The companionship bond may also be developed with some

FIGURE 18–1
Dogs and cats are the most popular companion animals, but animals such as ferrets are also popular.

unusual pets, such as pot-bellied pigs, reptiles, or fish. This wide range of pets provides companionship for people of all ages.

In recent years, cats have outnumbered dogs as the favorite pet. There are about 60 million cats and 52 million dogs in American households. Many cat owners have more than one cat, so although cats are more numerous, dogs are found in more households.

The popularity of dogs and cats as companions is due to a combination of factors. One of these is the ability of dogs and cats to form lasting social bonds with humans. They are able to establish a close relationship with their owners. They can also be house-trained fairly easily. House-training is impossible with many companion animals. Dogs and cats also have the advantage of being large enough for humans to interact with and play with and yet small enough to be kept in the house.

The most important reason for the popularity of dogs and cats is their ability to express themselves to their owners. Both dogs and cats have a number of gestures and sounds they use to communicate with their owners (Figure 18–2). People tend to

FIGURE 18–2
Dogs are able to express themselves to their owners.

FIGURE 18–3
Reptiles are rapidly becoming some of the more popular pets.

attribute human feelings and emotions to their pets because of this ability to be expressive.

The fastest-growing category of pets in the United States is reptiles (Figure 18–3). There are more than seven million pet reptiles living in homes in the United States. Iguanas are the most popular pet among reptiles. These lizards are cat-sized when full grown. They are clean, odorless, and can be house-trained. This category also includes snakes and tortoises, as well as amphibians such as frogs and salamanders.

Health Benefits

Pets are more than just companions—they are good for people's health. Scientists are now

Career Development Events

The FFA offers many opportunities for students to develop and display projects. Students even have the opportunity to develop their own programs and the area of companion animal production is well suited for this purpose. If you are involved with the training, grooming, care, or production of companion animals, you might want to organize an event that will showcase your animals.

The first step would be to locate others in your chapter with interest in companion animals. With the help of your advisor, form a committee to discuss the event. Begin by developing objectives and determining what you want to accomplish with the event. The objective might be to have a demonstration of grooming techniques, a competition in grooming, or a competition in the training of companion animals. Use your imagination—there are lots of types of events you could organize.

Once you determine what you want to accomplish, organize the procedures and establish guidelines to govern the event. Get everyone in your chapter involved in the planning and activities and with your advisor's help, organize the members into committees to get everything done. You might even get the local FFA Alumni Affiliate involved, which offers support for the FFA programs.

discovering that living with a pet contributes to the physical and mental well-being of humans. Relationships with animal companions appear to be beneficial to humans because they are uncomplicated. Animals are accepting and attentive, and respond to affection. They are not judgmental, they never talk back, they never criticize or give orders. Pets give people something to be responsible for and make them feel special and needed (Figure 18–4).

Companion animals are a good influence on children of all ages. They help children develop a sense of security. They have been used to encourage shy or withdrawn children to open up. Children who are normally hyperactive often become calmer around a companion animal.

FIGURE 18–4
Pets can make people feel needed.

As your event becomes successful, you may wish to invite other chapters to participate. This was the way most of the major FFA events began. The FFA is growing and looking for new events, and you just might have an idea that would grow into a national event.

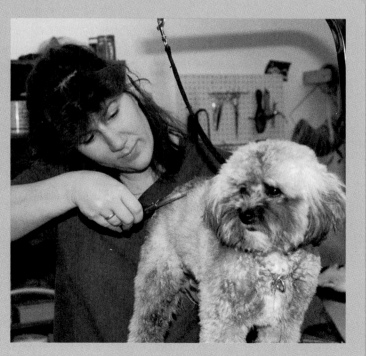

Students who raise companion animals as an SAEP may enter the Specialty Animal Production Proficiency Award.

FIGURE 18–5
Watching fish in an aquarium can help lower a person's blood pressure.

New evidence of the beneficial effects that companion animals have on human health is continually being discovered. We now understand that pets make people feel good, and a sick person who feels good mentally is likely to get better faster. People who have pets report fewer minor health problems, like colds and flu. Studies have also shown that petting a dog or a cat or watching fish in an aquarium can help lower a person's blood pressure and heart rate (Figure 18–5).

Due to the increasing mobility of today's society, many elderly people no longer have family members living close by. Therefore, companion animals play an important role in the lives of a number of older individuals. Many of them would feel isolated and alone without their pets. Pets provide them with companionship and a sense of being valued and needed. Older people may actually live longer, healthier lives because of their relationship with their companion animals.

Today about 50 percent of nursing homes use animals in some capacity. Bird aviaries and aquariums are popular among residents, as well as cats, rabbits, and guinea pigs. It is not recommended that dogs live in nursing homes full-time because they tend to be overfed by the residents, who cannot seem to resist feeding them cookies and treats. Today many volunteers take their dogs to visit the residents of nursing homes and hospitals in their community.

Service Animals

Some companion animals serve the dual roles of companionship and practical assistance. They are known as **service animals**. A number of agencies in the United States train assistance dogs. These dogs are trained to give more independence and mobility to people with disabilities. Dogs serve as the eyes, ears, or legs for thousands of people. Training usually takes between four and eight months, depending on the difficulty of the tasks that must be learned and the aptitude of the dog. Although training an assistance dog can cost thousands of dollars, many agencies provide them to people who need them at little or no cost.

Guide Dogs

The best known example of an assistance dog is the guide dog for the blind. The most commonly used breeds are German shepherds, Labradors, and retrievers (Figure 18–6). To work as a guide dog, the dog has to be exceptional. It has to be able to walk

FIGURE 18–6
The German shepherd is a popular breed used as guide dogs.

through crowds, climb stairs, ride on elevators, ride on buses and in cars. Most importantly, the dog has to be able to think for itself. It must learn to disobey a command if it could bring harm to its master.

Many of the organizations that train guide dogs have volunteers raise the puppies for the first year. Some of the volunteer puppy raisers belong to the 4-H organization. The volunteers are not expected to train the puppies as guides, but they are required to follow some basic rules. The puppies must be exposed to many of the activities that they will need to handle with ease as a guide. The puppy raisers are encouraged to take the puppies to the mall, to the park, to nursing homes, schools, or any other setting that involves the public. The puppies must be kept on a lead in public, and they must sleep next to the bed of the volunteer, just as they will when acting as guides.

The volunteers are warned not to play ball, tug of war, or other games with the puppies that could turn into bad habits when they become guide dogs. The puppy raisers must never feed their pets human food. Guide dogs must not be tempted by the smell or presence of human food, as they may accompany their owners to restaurants. They cannot be jumping up on tables begging for food.

Only about half of the puppies raised to be guide dogs successfully complete the training (Figure 18–7). During training, the dogs learn to work in a harness, and they learn commands like "Forward" and "Find the door". The dogs are trained through repetition and praise. They learn to ignore crowds, noises, squirrels, and cats. The dog must learn that it is as wide and as high as a person. It must be aware of obstructions that it can walk around or under, like awnings or branches that would impede its owner. A guide dog wearing a harness is on duty and should never by petted by other people. When

FIGURE 18–7
Only about half of the puppies raised to be guide dogs successfully complete the training. *Courtesy Guide Dog Foundation for the Blind, Inc.*

the dog is out of the harness, the dog is like any other family pet.

Hearing-ear dogs are trained to listen for people who cannot. Also called signal dogs, these animals can respond to more than 30 common household sounds like doorbells, telephones, alarm clocks, and fire alarms. They can even be trained to respond to a crying baby. The dogs alert their deaf or hard-of-hearing owners by walking back and forth from the source of the noise to the owner. Signal dogs are taught not to bark when alerting their master to a sound, as they would be unable to hear the barking. The dogs can also be trained to respond to sign language commands. Because the size of the dog is not important, hearing-ear dogs are usually mixed breeds and are often rescued from local animal shelters.

Service dogs are trained to help people who use wheelchairs or have spinal injuries. They are able to respond to more than 40 different commands. Service dogs can open doors, work light switches, pull emergency cords, and pull wheelchairs. Each dog may be trained a little differently in order to address the needs of the individual who will own the dog. Service dogs need to be large and are often retriever breeds.

Another companion animal that has been beneficial to humans is the horse. Horses are used in programs of physical therapy for people with disabilities. This type of therapy, called **hippotherapy**, can be used with people of all ages, but it is especially helpful for physically challenged children. The gait of a horse simulates the motion of humans as we walk. When we walk, our bodies move from side to side and up and down. Riding a horse recreates that sensation in people who are unable to walk unassisted (Figure 18–8).

Through hippotherapy, physically challenged individuals improve their balance, posture, strength,

FIGURE 18–8
Horses help people who are unable to walk unassisted. The gait of the horse resembles the motion humans make when they walk. *Courtesy of Cliff Ricketts.*

coordination, and muscle flexibility. In addition to the physical benefits, the riders also gain confidence. Hippotherapy can provide them with a whole new perspective and a sense of freedom. A horse can take its rider where no wheelchair could go. Volunteers are used in the care and maintenance of the horses as well as for lessons. Hippotherapy programs offer many opportunities to local agriculture programs and youth organizations.

S t o p

Diseases and Afflictions

Unfortunately, pets can give us more than companionship. Each year pets pass along infectious diseases to thousands of Americans. There are approximately 30 varieties of pet-borne illnesses, which are called **zoonoses**. Zoonoses are diseases and infections that can be transmitted from animals to humans. They can be passed on through direct contact with the animals or acquired indirectly through contact with animal feces or other contaminants. However, most diseases passed from animals to humans are easy to avoid and are treatable. Good hygiene and safe handling procedures should always be practiced when working with animals (Figure 18–9). Cats and dogs are responsible for the majority of zoonoses, but birds, fish, and turtles are also culprits.

Rabies is the best known and the most feared example of a zoonosis. It is contracted through the saliva of rabid animals. Although this disease is rare in pets, rabies is increasing in the wild animal population. Because pets may come into contact with wild animals, they should be vaccinated against the disease.

The common round worm of dogs is a **parasite** that can infect humans. A parasite is an organism that lives off of or at the expense of another organism. The parasite is transmitted through contact with the animal's feces or with contaminated soil.

FIGURE 18–9
Good hygiene and safe handling procedures should be practiced when working with animals.

FIGURE 18–10
Dog owners should make sure their pets are regularly wormed.

Children playing in areas frequented by dogs are especially at risk. Dog owners should make sure their pets are regularly dewormed (Figure 18–10).

Toxoplasmosis is a parasitic disease that can be transmitted by contact with cat feces. Toxoplasmosis can be especially dangerous to pregnant women, so most veterinarians recommend pregnant women not clean cat litter boxes.

Psittacosis, or parrot fever, is a disease transmitted by parrots, budgerigars, and other related caged birds. Humans can be infected by contact with the feces of contaminated birds. In handling birds and cages, dust masks or protective face shields should be used.

Ringworm is not a worm but a fungus that results in a skin aggravation in humans. It is primarily passed to humans by kittens and puppies. The animal appears unaffected because the fungus infects only the animal's fur. It is passed to you when you handle your pet. It is more common in children because adults seem to become more resistant to it with age.

Rocky Mountain spotted fever and Lyme disease are tick-transmitted diseases that can affect both humans and animals. Ticks are found in grassy, wooded areas, or they can be brought into the

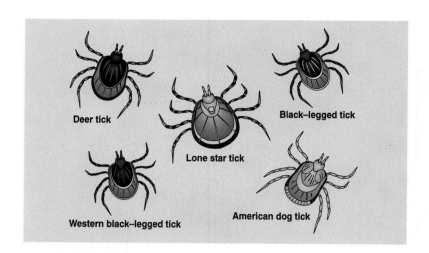

FIGURE 18–11
Ticks can transmit Lyme disease.

household by dogs and cats that have been outside (Figure 18–11). Both diseases are treatable with antibiotics. Normal grooming of animals after they have been outside will help locate and eliminate ticks.

Infections due to animal bites and scratches are another concern. The potential for infection varies. Less than 5 percent of dog bites become infected, but up to 50 percent of cat bites do. Cat scratch fever is associated with cat scratches or bites. This disease is not serious and can be treated with antibiotics. Safe handling techniques are an important measure to prevent bites and scratches, as well as injury to the animal. Prompt and thorough washing of pet bites and scratches with soap and water is always important.

Allergies are probably the most common afflictions that occur when humans have contact with animals. Many people develop allergies to animal hair and their dander, or flaking skin. Allergies cause hay-fever-like symptoms that can occur in both children and adults.

Industry

One measure of the value people place on pets is the amount of money they are willing to spend on them.

Americans spend over $20 billion a year on companion animals! Dog and cat food purchases account for about $8 billion of the total spent on pets. By comparison, people spend approximately $1 billion on baby food. The companion animal industry contributes to the economy through the sale of pets, pet food, accessories, and veterinary care (Figure 18–12).

The pet food industry utilizes many of the by-products and surpluses of the human food industry. The main ingredient in dry dog foods is grain, soybean meal, or wheat millings. The main ingredients in canned pet foods are meat by-products, which may include the waste products of meat that was processed for human consumption. The composition of pet food is carefully formulated to meet the nutritional needs of the animals. Canned, semi-moist, and dry foods are equally nutritious, but the canned varieties generally contain a higher percentage of protein and fat. Consequently, pets should be fed smaller amounts of canned food or mix dry foods with canned.

Cats have special dietary needs. Pet foods formulated for cats have a much higher protein and fat content than any type of dog food (Figure 18–13). Dogs often seem to prefer cat food because of the high fat content. For healthy, active dogs, nibbling a little cat food is not a problem, but they will become obese if allowed to eat it regularly. In contrast, cats should not be allowed to eat dog food because it does not contain all of the nutrients that a cat needs to stay healthy.

The supply of pet accessories is also an important contributor to the economy. A companion animal may need any number of essential items such as a feeding bowl, dish, or trough; a cage, hutch, or kennel; and chew bones, scratch poles, and toys. Many pet owners also indulge in luxury items such as heated pet beds, jewelry, pet clothing, and car seats.

FIGURE 18–12
The pet food industry produces a wide variety of pet foods.

FIGURE 18–13
Cats require a diet higher in protein and fat than dogs. *Courtesy of James Strawser, The University of Georgia.*

Health Care

Americans spend over $10 billion every year on the health care of their companion animals. There are approximately 40,000 veterinarians in the United States; one-third of them treat small animals exclusively (Figure 18–14). Like all veterinarians, those who specialize in the care of companion animals perform a wide variety of tasks every day. They treat animal injuries, set broken bones, immunize healthy animals against disease, and perform surgery.

Veterinarians today can successfully perform hip replacements and kidney transplants on companion animals. They have also performed balloon angioplasty to open clogged arteries, open-heart surgery, and dental surgery. In recent years, medical care for animals has become as highly technical as medical care for humans. In fact, new medical procedures are perfected on animals, so in some instances animals may get even more advanced care than humans. However, the costs of these health care advances may be more than some pet owners are willing or able to bear.

Veterinarians stress preventive measures when they counsel pet owners. They encourage the pet

FIGURE 18–14
One third of all veterinarians in the United States treat small animals exclusively.

owner to include vaccination programs and regular dental exams in their pet's health care plan. According to veterinarians, one of the most common problems they see in dogs and cats today is obesity. Half of American dogs and almost one-fourth of cats are obese. The animals suffer from overfeeding and a lack of exercise. Some animals simply do not get enough attention. When animals get bored, they have a tendency to eat too much, just like people. A sound diet and daily exercise routine should also be a part of the pet's overall health care plan.

Responsible Ownership

Responsible ownership of a companion animal requires a long-term commitment and a realistic assessment of the needs of the pet. A number of factors should be considered by the prospective pet owner that will contribute to the well-being of the pet. The first is the amount and kind of space in which the pet will live. A lively hunting dog does not belong in an apartment (Figure 18–15). Likewise, an easily frightened pet should not be kept in a household with several lively, noisy children. No dog should be tied to a tree and left alone all day, every day. Animals that are forced to live in environments

FIGURE 18–15
A lively hunting dog does not belong in an apartment.

FIGURE 18–16
Small animals such as mice do not require much space.

that do not suit their natures may exhibit behavioral problems.

Although many households want to have dogs, busy families that do not have the time or space to devote to a dog have many other options to consider. Rabbits, hamsters, and guinea pigs can be affectionate, cuddly pets and do not require the time or the space of a dog (Figure 18–16). Reptiles, such as geckos, monitors, and iguanas, are not soft and cuddly, but many of them do seem to enjoy contact with people, probably because of human body heat. Reptiles do not require much space and may even be kept in apartments that do not allow cats and dogs.

Responsible owners of companion animals also need to consider the problem of pet overpopulation. Millions of unwanted and homeless animals are born each year. This overpopulation creates problems for society and leads to suffering for many companion animals. To solve this problem, the number of puppies and kittens born must be reduced.

At least 10 million dogs and cats are destroyed, or euthanized, each year by animal pounds and shelters. Future pet owners are encouraged to consider rescuing an animal by adopting a pet from their local animal shelter. It is the responsibility of current pet owners to help control this problem by neutering and spaying their pets.

Spaying and neutering are surgical procedures that prevent animals from reproducing. Spaying is performed on females and neutering is performed on males. The animals are anesthetized so they do not feel any pain during the surgery. They may experience some discomfort for a day or two after the procedure.

Spaying and neutering a pet can result in a healthier, happier animal. Spayed females are less susceptible to some infections and cancerous growths. Neutered males are less likely to display aggressive behaviors or to roam.

Summary

Companion animals represent a large and growing segment of animal production. The many segments of this vast industry have a positive effect on the economy. It is different from most of the agricultural industry in that it is not involved with the production of food and fiber. It does, however, provide a valuable product for people who enjoy animals as companions.

18 CHAPTER REVIEW

Student Learning Activities

1 Visit the local pet store and interview the owner as to where the animals come from. Research how the animals are produced.

2 Make a list of all the money spent on your pet during the last year. Determine how many different industries were supported by the expenditures.

③ Visit the local humane society and determine how your class can help in educating people about animal care.

④ Go to the Internet and research a type of companion animal. Determine how long it has been in domestication and the country of origin. What are some of the popular breeds?

True/False

① ___ There are more cats than dogs in the United States.

② ___ Dog food can also be fed to cats with no ill effects.

③ ___ A companion animal may be a snake.

④ ___ A zoonosis is a disease that is passed from a caged animal to a companion animal.

⑤ ___ Lyme disease is contracted through contact with cat feces.

⑥ ___ Rabies is relatively uncommon in household pets.

⑦ ___ Veterinary technology tends to lag far behind human medical technology.

⑧ ___ Most dog bites become infected.

⑨ ___ The fastest growing category of pets in the United States is reptiles.

⑩ ___ Spaying is a surgical procedure performed on female animals.

Multiple Choice

① Ringworm is caused by
    a. parasites
    b. fungus
    c. worms

② Signal dogs are trained to assist
    a. people in wheelchairs
    b. the blind
    c. the deaf

3 The main ingredient in dry dog food is
    a. grain
    b. chicken
    c. meat by-products

4 Hippotherapy is particularly beneficial for
    a. the elderly
    b. physically challenged children
    c. the deaf and hard-of-hearing

5 Volunteers who raise puppies for guide dog programs should try to expose the puppy to
    a. games like tug-of-war
    b. human food
    c. crowds of people

6 Neutering is a surgical procedure that is performed on
    a. male animals
    b. reptiles
    c. female animals

7 Service dogs are trained to help
    a. the blind
    b. the sick
    c. people in wheelchairs

8 The best-known and most feared zoonosis is
    a. rabies
    b. Lyme disease
    c. ringworm

9 It is not recommended that dogs live in nursing homes full-time because
    a. they tend to be aggressive toward elderly people
    b. they tend to be overfed by the residents
    c. companion animals are not beneficial for elderly people

10 Psittacosis is transmitted by
    a. contact with the feces of some caged birds
    b. ticks
    c. parasites

Discussion

1. Why is spaying or neutering a pet an important part of responsible pet ownership?

2. How does the companion animal industry contribute to the economy?

3. What are the differences between dry and canned pet foods?

4. Why are dogs and cats the most popular companion animals?

5. What is a companion animal?

6. How do companion animals affect the health of humans?

7. Why are companion animals beneficial to elderly people?

8. What are zoonoses?

9. How do assistance dogs help people?

10. What is hippotherapy?

CHAPTER (19)

Preserving Our Food Supply

Student Objectives

When you have finished studying this chapter, you should be able to:

- Discuss the need for food preservation.
- List the major ways food is preserved.
- Discuss modern trends in food consumption.
- Explain how food can become unsafe to eat.
- Discuss how the USDA helps to ensure a safe food supply.

Key Terms

drying	canning	USDA
jerky	blanching	pesticide residues
pickling	freeze-drying	E. coli bacteria
salting	Mad Cow Disease	

One of the most important segments of our huge agricultural industry is concerned with preserving our food. No matter how much food we produce, it is of little use if it does not reach consumers in an edible form. As you know from studying the science of living organisms, plants and animals begin to decompose when they are no longer living. Microorganisms such as bacteria feed on plant and animal tissue as a means of returning dead organisms back to the soil. This happens with our food crops also. When plants and animals are harvested for food, the tissues within the food begin to break down. Without a means of preserving the food, it would spoil before it could be eaten.

While some of our food is sold fresh, the vast majority of food products are preserved in some manner. Many of our green leafy vegetables are sold fresh, but these foods do not last long in the fresh state. Fruits such as apples, bananas, and grapes are bought fresh but still need refrigeration to make them last. Also, meat such as beef, pork, and chicken is sold fresh, but it must be eaten within a few days or it will spoil. The percentages of these foods that are sold fresh keep declining. In our modern era,

FIGURE 19–1
Many of our green leafy vegetables are sold fresh, but these foods do not last long in the fresh state. *Courtesy of USDA.*

FIGURE 19–2
With the advent of the microwave oven, there is great demand for food that can be taken from the freezer and cooked in a very few minutes.

most households have do not have someone whose sole job is that of keeping house, shopping, and cooking. People tend to want to visit the grocery store less often than in years past. This means that food must last longer than just a few days. Another trend is toward foods that are not only frozen but are also cooked. With the advent of the microwave oven, there is great demand for food that can be taken from the freezer and cooked in a very few minutes.

Methods of Preserving Foods

Before methods of preserving foods were developed, people feasted when food supplies were plentiful. For example, when a large animal was killed, there might be plenty of meat for everyone. However, this only lasted a few days, because the meat began to spoil. When fruits were ripe or vegetables were at their peak, people ate well. During the colder months, when the fruits and vegetables disappeared, there was none available. Another problem encountered was that of travel. When people traveled, they most often had to take food with them. If the trip was long, fresh food would spoil before the journey ended. This meant that food needed to be preserved to last the entire journey.

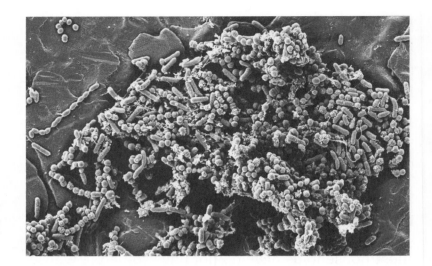

FIGURE 19-3
Bacteria such as this cause food to spoil. *Courtesy of Agricultural Research Service.*

Perhaps the greatest problem was that of feeding an army, especially if that army traveled. Not only did food have to be preserved, it also had to be preserved so that it would taste good when it was eaten. These problems led to the development of food preservation. Basically, all food preservation methods have at least one thing in common—they slow down or stop the growth and reproduction of microorganisms such as bacteria.

Drying

The first method of food preservation was probably **drying.** People noticed that as plant seeds matured, they became dry. This allowed the seeds to survive during the winter months. People began to use the sun and fires to slowly dry foods. Fruits were dried and preserved. Meats were cut into thin slices and hung over a fire until it dried. Many Native Americans in the Pacific Northwest survived by catching and drying salmon that were caught during certain time of the year when the fish came up small streams.

Drying preserves food because the moisture is removed from the plant or animal tissues. In order

FIGURE 19–4
Drying preserves such foods as raisins. Note the difference between the raisins after drying and the grapes before drying. *Courtesy of Agricultural Research Service.*

to survive, microorganisms must have moisture. Without moisture, they cannot grow and reproduce. This method worked so well that it has survived from prerecorded history until the present. We still use this method to preserve food. Dried fruits such as dates, prunes, and raisins are readily available in the grocery store. Dried meats such as **jerky** can be found in most convenience stores.

Pickling and Salting

Even though this method of food preservation goes back hundreds of years, it is still used today. Almost all of us enjoy pickles on our hamburgers and sour kraut on our hotdogs. **Pickling** preserved both of these foods. Vinegar has been used as a medium for pickling for centuries.

It works because vinegar is so acid that bacteria will not live in the liquid. Sauerkraut is preserved because of the addition of salt and because the cabbage undergoes a fermentation process very much like farmers use to create silage for livestock. This process also makes the sauerkraut more acid and this also aids in preservation.

FIGURE 19–5
Pickling preserves food by using vinegar as a medium.

Like pickling, salt has been used to preserve food for almost as long as we have recorded history. High concentrations of salt cause bacteria to die by drawing the moisture out of organisms. This process is called osmosis. Meats and fish have been preserved by this method, which is called **salting.** In the early days of our country, settlers used salt to cure their meat. Buildings known as smoke houses were constructed to hang the meat. Salt was forced into the meat and it was smoked to prevent insect damage. Even today, salt cured hams and fish are very popular.

Canning

We all use cans of fruit, vegetables, meats, and a wide variety of prepared foods. This process works by placing foods in metal cans or glass jars. The cans or jars are then sealed so that nothing can get in or get out until the container is opened. Subjecting the filled containers to a high level of heat kills the bacteria and other organisms. This process, called **canning,** is done in what is called a retort. This device puts the cans or jars of food under pressure supplied by steam. Most food contains water. If the cans of food were heated high enough to kill all of the microorganisms, the water

FIGURE 19–6
Hams are preserved by salting and smoking. This food is popular all over the world.

in the cans would turn to steam and build enough pressure to explode. As you know from science class, water evaporates (turns to vapor) at 212 degrees F at atmospheric pressure. However, if pressure is added, the temperature needed to change water to vapor (steam) increases. Putting the containers of food under pressure allows them to reach high enough temperature to kill bacteria and other organisms in the food.

Refrigeration

The invention of refrigeration brought about a revolution in the way foods are preserved. When the temperature is lowered, microorganisms cannot reproduce as rapidly as they can at warmer temperatures. Fresh fruits and vegetables as well as all types of meat can be preserved for days under refrigeration.

This means that these food products can be shipped all across the country on refrigerated trucks or rail cars. Oranges grown in Florida can be shipped all the way to Canada and beef slaughtered in Iowa can be shipped to both coasts. Because of refrigeration, we can enjoy a much broader array of different foods.

FIGURE 19–7
In the canning process, food is sealed in a can and the food is sterilized using heat. *Courtesy of Getty Images.*

#5

FIGURE 19–8
Meat can be preserved for days using refrigeration. *Courtesy of USDA.*

Freezing

A common way of preserving food is by freezing. While lower temperatures (refrigeration) slows the growth of microorganisms, very low temperatures (below 0 degrees F) keep bacteria from growing and reproducing, so the food does not spoil. Fruits, meats, vegetable, fish and other forms of food are harvested, processed, and quickly frozen. Some foods such as vegetables undergo a process called **blanching.** This procedure involves immersing the vegetables in hot water then rapidly submerging them in ice water. This stops enzymes in the vegetables from causing changes in the plant tissues. Enzymes are substances that cause or speed up a chemical reaction.

Career Development Events

The FFA offers students many opportunities to develop and display projects. The National FFA Agriscience Fair recognizes students who are studying the application of scientific principles and emerging technologies in agriculture. Much like that of the International Science and Engineering Fair, the Agriscience Fair reflects an agricultural theme. Participation begins at the local chapter level and advances to state and national levels. Competition is open to all FFA members in grades 7–12.

One of the major goals of the Agriscience Fair is to provide students with an opportunity to use the scientific process. They can demonstrate and display Agriscience projects while reinforcing skills and principles learned in Agriscience courses. Award winners are also recognized for their efforts. Certificates, medals, ribbons, plaques, and even scholarships and cash awards may be given to division winners in each category.

There are five categories for the National FFA Agriscience Fair: Biochemistry/Microbiology/Food Science, Environmental Sciences, Zoology (Animal Science), Botany (Plant/Soil Science), and Engineering (Mechanical/Agriculture Engineering Science). Each member or team of two members may enter only one project. The participants are

Once the food is frozen, bacteria cannot reproduce in the low temperature. When the food is thawed, the bacteria can then begin to reproduce and the food will spoil unless it is prepared and eaten within a short period. This process also allows food to be cooked and then frozen. Microwave ovens are often used to turn precooked frozen foods into ready to eat meals. This is becoming an increasingly more popular way to buy and prepare meals.

A variation of freezing foods is the process of **freeze-drying** foods. This method combines the concepts of both freezing and drying. Remember from your science class that water comes in three states—liquid, gas, and solid. At room temperature it is a liquid. At 32 degrees F, it begins to turn to a solid (ice). At 212 degrees F, it begins to turn to a gas (vapor

THE EFFECT OF A NICKEL ACCUMULATING PLANT ON ANIMAL BROWSING

A major goal of the Agriscience Fair is to provide students with an opportunity to use the scientific process.

required to meet with the judges to explain their project and will be judged on how well the scientific method was followed, the detail and accuracy of the log book and project report, and whether tools/equipment were used in the best possible way. Other categories that are scored include knowledge gained, information, thoroughness, conclusions, interview, and visual display. For more information concerning the National Agriscience Fair, contact your agriculture teacher or visit http://www.ffa.org.

FIGURE 19–9
Once food is frozen, bacteria cannot reproduce. *Courtesy of USDA.*

FIGURE 19–10
Freeze dried foods are well preserved and light weight. *Courtesy of Comstock Images.*

or steam). Also remember that these changes take place at these temperatures at atmospheric pressure. If water is put under pressure it has to get much hotter to turn to vapor. If pressure is removed in a vacuum, water will turn to vapor at a lower temperature. In the freeze-drying process, frozen food is placed under a tight vacuum. Even though it is frozen, the water in the food will evaporate, even at temperatures below 0 degrees F. Another way to explain the process is that it is the reverse of the principle used in canning. This process results in a well-preserved food product that weighs relatively little. Back packers use freeze-dried foods extensively.

Stop!

Food Safety

Americans enjoy the most abundant, nutritional and safest food supply of any people in history. Almost all consumers are concerned about the food they eat. After all, our health and well-being depends on the wholesomeness of the food we eat. However, some people are overly concerned about the safety of our food supply. This concern comes about as a result of instances where food has caused health problems.

In recent years, people have become sick and some have died from eating food at fast food restaurants. There is also the scare over **Mad Cow Disease** that was discovered in an animal scheduled for slaughter. This disease occurs in cattle. A type of protein called a prion that attacks the nervous system causes it. Mad Cow Disease can be transmitted to humans if they eat poorly cooked beef from an animal with the disease. It is almost always fatal.

While all these are real concerns, the overwhelming majority of our food is safe and wholesome. Generally all problems with food can be traced to one of three areas: food contamination, food spoilage, or improper food preparation. Food may be contaminated in a variety of ways. For example insects may have gotten into the food product. Insects can carry a variety of microorganisms that transmit disease. Another way that food is contaminated is by pesticide residuals. Also, foreign particles can be accidentally added to food during processing. Anything that comes in contact with food that interferes with the purity of the food product causes contamination. Food contamination is generally considered to be a relatively minor problem because it happens so infrequently. Government agencies under the direction of the United States Department of Agriculture **(USDA)** monitor all food products are processed, preserved, packaged. Food products are constantly inspected for contamination of any kind.

Perhaps the most closely monitored of all food products is that of meats. Trained meat inspection experts carefully inspect each and every animal that is slaughtered. The inspectors look for signs of animal diseases or anything else that might compromise the purity of the meat product. All foods are monitored for **pesticide residues.** Almost all crops and animals are subjected to pesticides at one time or another. The USDA restricts the type of pesticide

FIGURE 19–11
The overwhelming majority of our food products are safe and wholesome. *Courtesy of USDA.*

FIGURE 19–12
Meat products are closely monitored and inspected. *Courtesy of USDA.*

that can be used on food plants and animals. The amount of pesticide on food products is also checked to ensure that whatever residues left on the products are not harmful.

A common contaminant on meat is **E. coli bacteria.** Sometimes during the slaughtering process, bacteria from animal feces come in contact with the meat. This bacteria can make us ill if it is consumed with the meat. This means that meat must be thoroughly cooked to kill any bacteria on the meat. This is particularly true of ground beef that we eat as hamburgers. Because the meat is ground up into many tiny particles, there is a lot of surface area that may be contaminated with E. coli bacteria. Meats such as steaks and roasts should be cooked until the center of the meat reaches at least 145 degrees F. Ground beef should reach a temperature of at least 165 degrees F and poultry should be cooked to 180 degrees F.

Care should also be taken in the way meat is prepared for cooking. If the meat is frozen, it should be thawed in the refrigerator or in a microwave oven. It should never be thawed at room temperature. Thawing at room temperature allows the rapid

FIGURE 19–13
To ensure safety, poultry should
be cooked to 180 degrees F.
Courtesy of Getty Images.

buildup of bacteria because of the warm temperature and the time it takes to thaw. Dishes, counter-
tops, and utensils that come in contact with raw
meat should be thoroughly washed in hot soapy
water. Also, hands should be washed after handling
meat.

Foods can also spoil after they are packaged. In
fact, all food products will eventually spoil over
time. Frozen foods, canned foods, dried foods, and
pickled foods will sooner or later become inedible.
Preserved food products have an expiration date
printed on the package that indicates the last day
the product can be sold.

For example, a jug of milk may have a date printed on it. This does not mean that the milk is spoiled
after the day indicated. This means that the grocer
cannot sell the milk after that day. Sometimes the
label designates the day when food loses its fresh-
ness. A box of cereal may have such a date. The cereal may be safely eaten after that day, but it might not
taste as fresh.

Rarely a food may spoil before the expected time.
This is usually because of improper preservation
or packaging. Food that spoils in this manner is

FIGURE 19–14
Preserved food products have an expiration date printed on the package that indicates the last day the product can be sold. *Courtesy of USDA.*

#10

dangerous. The same bacteria that cause the spoilage can also make us ill. Beware of any food package that seems swollen or cans of food that have a domed top. This is an indication that bacteria have rapidly reproduced. These bacteria produce wastes and gases that cause the sealed can to expand and bulge out the top. These cans should be thrown away and not opened. Spoiled food may also have an unpleasant odor or off-flavor. If you notice a bad odor or bad taste in food, do not eat it. Any spoiled food should be disposed of immediately.

Summary

American farmers produce an abundance of food for our population. Modern techniques allow us to preserve the food so it can be eaten year-round. Regulations by the USDA, proper storage methods, and proper cooking practices allow us to enjoy the safest, most wholesome food supply in the world. Following proper procedures when preparing foods and watching for danger signs will help keep our food safe.

19 CHAPTER REVIEW

Student Learning Activities

1. Make a list of all the food in your home. Beside the name of each food, write down the method that was used to preserve the food. Which method do you think best preserves the food's flavor? Which of the methods are the easiest to prepare for cooking?

2. Interview an elderly person about how food was preserved when he/she was young. Have there been any changes? What do you think brought about the changes?

3. Visit a grocery store and make a list of all the different ways meat is preserved. Why are there so many different ways? Which method do you like the best? Why?

4. Conduct an Internet search and locate information on new techniques of preserving food. Which do you think has promise? What are some of the problems? Do you think these problems can be solved?

True/False

1. ___ Almost all of the food we eat is sold fresh.

2. ___ People visit the grocery store less often than in the past.

3. ___ The first method of preserving food was probably drying.

4. ___ Dried foods are no longer available.

5. ___ The use of pickling as a method of preserving food is only about 20 years old.

6. ___ Salt cured hams are popular today.

7. ___ Refrigeration kills all of the bacteria in food.

8. ___ Americans have the safest most wholesome food supply in history.

⑨ ___ Rare hamburgers are tastier than well cooked hamburgers and should be eaten often.

⑩ ___ A can of food with a domed top is a sign of food spoilage and should not be opened.

Multiple Choice

① A dried meat product is known as
    a. potted meat.
    b. sausage.
    c. jerky.

② A common substance used in pickling food is
    a. vinegar.
    b. alcohol.
    c. olive oil.

③ A device used to put cans of food under high pressure and heat is called a
    a. vacuum chamber.
    b. retort.
    c. cooker.

④ Blanching is a process that kills
    a. bad taste.
    b. microorganisms.
    c. enzymes.

⑤ An advantage of freeze dried foods are that they are
    a. lighter in weight
    b. less expensive
    c. easier to process

⑥ Mad Cow Disease is caused by a
    a. bacterium.
    b. prion.
    c. virus.

⑦ E. coli bacteria is a big problem with
    a. vegetables.
    b. fruits.
    c. meats.

8 Poultry should be cooked to a temperature of at least
    a. 125 degrees F.
    b. 460 degrees F.
    c. 180 degrees F.

9 The expiration date on milk means that
    a. the milk spoils on that day.
    b. the milk cannot be sold after that day.
    c. has no real meaning.

10 Spoiled food can be detected by
    a. a bad taste.
    b. a bad smell.
    c. both a and b.

Discussion

1 What causes food to spoil?

2 Why do people visit the grocery store less often than in the past?

3 What problems did the invention of food preservation solve?

4 List some modern foods that are preserved by drying.

5 Explain why the invention of refrigeration was so important to the food industry.

6 Why is it important to put frozen food under a vacuum in the freeze drying process?

7 List three ways food can become contaminated.

8 Explain why it is so important to thoroughly cook meat.

9 Where does *E. coli* bacteria come from?

10 What causes a can of food to have a domed top?

CHAPTER **20**

The Ethical Treatment of Animals

Student Objectives

When you have finished studying this chapter, you should be able to:

- Distinguish between arguments of animal rights and animal welfare proponents.
- Explain why producers treat their animals well.
- Discuss why people may feel that agricultural animals are mistreated.
- Give reasons why confinement operations are not cruel to animals.
- Discuss why the use of management practices is humane.
- Discuss the use of animals for experimentation.

Key Terms

animal welfare   confinement operations   farrowing crates

animal rights   farm factories

FIGURE 20–1
Today relatively few people really understand how agricultural animals are raised. *Courtesy of James Strawser, The University of Georgia.*

People have always used animals. Since humans first appeared on the Earth, they have hunted and killed animals for food, shelter, and clothing. Very early they began to raise animals in addition to hunting them. From the earliest recorded history to today people have controlled the lives of animals they raised.

In years past, most people were familiar with farm animals and understood how they were raised and sold for slaughter. Today in the United States, only a small percentage of the people really understand about the raising of farm animals (Figure 20–1). This creates an atmosphere where much of what happens can be misunderstood. Some people see the raising and management of animals to be cruel. They feel it is unjust to grow and slaughter animals for our use.

There are at least two lines of thinking concerning the well-being of animals. The first is generally referred to as those who endorse **animal welfare.** These people feel that it is all right to raise animals for human use. They say that animals should be cared for properly and that the animals should be comfortable in their surroundings. The other group, referred to as **animal rights** advocates, feel that animals should be free and that they have as much right to live as do humans. They uphold the notion that all animals that are kept in pens, cages, or pastures should be set free.

Animal Welfare

Producers of agricultural animals would be the first to agree that animals should be treated well. The problem comes about when animal welfare activists disagree with producers over the question of what is proper treatment. Producers know that animals have to be cared for properly. Animals that are abused or mistreated in any way will not grow or produce very well. Producers raise animals for profit. Animals that are miserable in their surroundings do not eat, grow, and reproduce efficiently. Unless animals grow fast and reproduce, the producers cannot make much profit from the animals. Because of this, it is in the best interest of the producer to treat the animals well (Figure 20–2). In addition, most producers raise animals for a living because they like animals. For thousands of years, humans have almost literally lived with their animals to care for and protect them. Sheepherders have always had a reputation for caring for their flocks, often at the risk of their lives (Figure 20–3).

Millions of dollars and countless hours of time have gone into research to determine what makes animals comfortable. Modern facilities have been designed to make animals as comfortable as possible. Housing is built so that the space for the animals is not too warm or too cold. Producers realize that if animals are too hot or too cold, they are stressed. Stress causes animals to use energy that could otherwise be used for putting on weight.

FIGURE 20–2
Producers know that to make a profit animals have to be well cared for. *Courtesy of James Strawser, The University of Georgia.*

FIGURE 20–3
Sheepherders have always had a reputation for providing complete care for their flocks. *Courtesy of James Strawser, The University of Georgia.*

Confining Animals

Animal welfare proponents sometimes object to putting animals into **confinement operations.** They say it is wrong to give animals only a limited space to live (Figure 20–4). They point out examples of pigs being raised their entire lives in small pens, laying hens being confined to cages, and sows being kept in farrowing crates. The animal welfare activists

FIGURE 20–4
Some people feel it is cruel to raise animals in a confinement operation. *Courtesy of James Strawser, The University of Georgia.*

contend that animals are being mass produced in what they call corporate-owned **farm factories.**

Producers point out that most of the animals raised are produced on family farms and not on corporate-owned farms. They also argue that facilities are designed specifically for certain animals. For example, a hog house is designed for hogs. This design takes into account the amount of space that a hog needs to be comfortable. It also accounts for the nature of the animal. Pigs are sociable animals and like to be around other pigs. To separate them into isolated spaces where they are by themselves would stress the animals. Research has shown that most animals do better when raised with other animals of their own kind.

The use of **farrowing crates** is another example of a facility designed and used for the betterment of the animal. To those who do not understand the purpose and use of farrowing crates, it might look cruel. Sows are placed in the crates about a week before they give birth. They remain in the crate for about a week after the pigs are born. The sow cannot turn around in the crate and has only room to lie down. The purpose is to prevent her from crushing her piglets as she lies down (Figure 20–5).

FIGURE 20–5
Sows are kept in farrowing crates to protect the baby pigs. *Courtesy of James Strawser, The University of Georgia.*

The crate is designed so that the sow has to lie straight down instead of flopping onto her side. This slower movement allows the little pigs time to move before the mother lies on them. The sows are given plenty of feed and water and the area is kept dry and warm. The sows seem to be in no discomfort and many young pigs are prevented from being crushed to death.

Animals have been selectively bred to live in confinement. This means that they are quite different from their ancestors that roamed freely. Layers that are kept in cages would probably not do well roaming free. The cages are designed for cleanliness and comfort for the hens. They have an abundant source of feed and water and the temperature in the laying houses is carefully controlled (Figure 20–6).

Animals that are kept in confinement can receive better care than animals that roam free. Parasites have always been a problem to animals. They cause stress and discomfort and may even shorten the life of the animal. When animals are kept in confinement, producers make certain that the animals are kept free of both internal and external parasites. In addition, the animals do not have to worry about predators like animals in the wild or those running free. The producers see the animals every day and care for their needs.

Management Practices

Another problem perceived by animal welfare activists is that of the management practices used by producers. They say that practices such as branding, castration, docking, and dehorning are cruel because they put the animal through a lot of pain (Figure 20–7).

Producers point out that these practices are done for the benefit of the animal. For example, the removal of horns from an animal may cause temporary pain during and after the operation.

FIGURE 20–6
Hens that are kept in cages have been bred for this type of life. *Courtesy of James Strawser, The University of Georgia.*

FIGURE 20–7
Management operations are often viewed as being cruel. *Courtesy of Cooperative Extension Service, The University of Georgia.*

Career Development Events

You already know that agriculture is more than farming. Any students interested in a career in livestock should strengthen their knowledge and experience by becoming active in the FFA. The FFA develops competent and aggressive leaders who learn how to speak out and debate issues such as animal welfare. Agriculture education and the FFA provide hands-on experience in practicing skills of health management with animals. This experience gives you a better understanding of animal welfare issues.

Animal welfare has become an issue of debate in local, state, and national agricultural meetings around the world. To speak out and present your views requires the use of correct parliamentary procedure. All FFA members should learn how to conduct and participate in local meetings. Parliamentary procedure teaches the student an orderly method of conducting business that assures that all sides of an issue are treated fairly. Most FFA chapters conduct a local Parliamentary Procedure Event where six students make up a team. The team is given tasks to perform using correct parliamentary procedure. In preparing for the event, team members must learn proper use of the gavel, voting procedures, order of motions, and other requirements for properly conducting a meeting. Team members must individually answer questions covering basic parliamentary law and perform as a team. Parliamentary procedure is an effective

FIGURE 20–8
Domestic animals have no need for horns. *Courtesy of James Strawser, The University of Georgia.*

However, the purpose of horns is for self-defense. Animals under the care and protection of humans have no need for horns (Figure 20–8). During mating or other times the animals may use the horns on each other. If they are removed at an early age, there is far less likelihood they will injure each other.

Castration will also prevent the animals from fighting. Neutering makes animals more docile and less aggressive. Normally, male animals fight to establish mating dominance. Castration prevents this behavior and prevents injury both to animals and humans. Practices such as docking (removing the tail) help keep the animals clean. Remember

method of addressing issues while respecting the vote of the majority and the rights of the minority. The chapter winning team will have the opportunity to compete in area, state, and national FFA Parliamentary Procedure events.

FFA Activities such as public speaking help prepare students to address issues such as animal welfare. *Courtesy of Carol Duval, National FFA Organization.*

that all the practices used by producers are intended to help and not harm the animal. It is in the best interests of the producer and the animal to keep the animal healthy, comfortable, and contented.

Some practices, such as hot-iron branding, are used less now than in the past. Newer techniques, such as freeze branding and tattoos, are being used. These marking techniques cause less stress to the animals.

Several years ago, the National Cattlemen's Association adopted the following statement of principles on animal care, environmental stewardship, and food safety:

I believe in the humane treatment of farm animals and in continued stewardship of all natural resources.

I believe my cattle will be healthier and more productive when good husbandry practices are used.

I believe that my and future generations will benefit from my ability to sustain and conserve natural resources.

I will support research efforts directed toward more efficient production of a wholesome food supply.

I believe it is my responsibility to produce a safe and wholesome product.

I believe it is the purpose of food animals to serve mankind and it is the responsibility of all human beings to care for animals in their care.

Similarly, the National Pork Producers Council has adopted the following Pork Producers' Creed:

I believe in the kind and humane treatment of farm animals and that the most efficient production practices are those that are designed to provide comfort.

I believe my livestock operation will be more efficient and profitable if managed in a manner consistent with good husbandry practices as known and recommended by the animal husbandry community.

I believe in an open door policy to visitors to my farm, to all those who are sincerely interested in production methods and the welfare of animals, so long as they do not endanger the health and welfare of my animals and do not interrupt my production routine or impair the production process.

I believe in and will support research efforts designed to measure stress of farm animals and directed toward more efficient production of food and enhancement of the welfare of animals and man.

I believe it is the animal's purpose to serve man; it is man's responsibility to care for the animals in his charge. I will vigorously oppose any legislation or regulatory activity that states or implies interference with that responsibility.

Experimentation on Animals

Another issue of concern for animal welfare activists is the use of animals for research (Figure 20–9). For many years, animals have been used to test products and procedures for humans. For example, before any drug is tested on humans, it must be first tested on animals. Medical procedures such as operations and vaccinations are tested on animals. Most of the medical advances made by humans

FIGURE 20–9
Some people object to using animals for experimentation.

during the past century have come about through the use of animals to test treatments and medication. Research scientists argue that the use of experimentation on animals is well justified by these advances. They cite the examples of diseases, such as polio, that have been almost eradicated through the use of research using animals in experiments.

Most people realize that research is carried out using mice, rabbits, and guinea pigs. However, during recent years much controversy has come about over the use of cats, dogs, and primates. Some people feel that it is worse to use animals that are "of a higher order." Those who oppose the use of animals in research contend that the knowledge gained through the research cannot justify the suffering the animals must undergo. They advocate the use of computer simulations to test treatments. Researchers point out the computer simulations are helpful, but nothing can ever replace real animals.

Research scientists have a much more difficult time justifying the use of animals to test such products as cosmetics. Activists insist that the suffering of animals should not be allowed merely to produce new products that are used only to make people appear more attractive.

Animal Rights Activists

Another group concerned with the treatment of animals is animal rights activists. They contend that animals have as much right to enjoy life and to be free as do humans. They believe that killing an animal is just as wrong as killing a human. They oppose the raising of animals and the use of animal products. They also strongly oppose hunting and feel that animals should be free to live in the wild.

Opponents to animal rights activists contend that animals are not the equal of humans and that they exist for use by humans. They argue that if

FIGURE 20–10
Animals that have been bred to be cared for by humans could not live on their own. *Courtesy of James McNitt.*

animals were turned loose, both animals and humans would suffer. Animals that have been bred to be cared for by humans could not care for themselves (Figure 20–10). If animals were turned loose without care they would lead short, miserable lives. Domesticated animals live the best lives under the care of humans.

Opponents also point out that the loss of animal products would be a burden to humans. Nutrients obtained from the eating of meat would be difficult to replace from other sources. The absence of drugs and other pharmaceuticals made from animal by-products would create problems for those people whose lives depend on these products.

Animals left to live on their own would have a difficult time reproducing and surviving. When left entirely on their own in a closed ecological system, animal populations might be able to achieve a balance. However, human populations are large enough to prevent this from happening. A good example is that of deer populations that are allowed to grow entirely on their own. Unless the population is kept under control by measures such as hunting, there are too many of the animals (Figure 20–11). When this happens, the herds of animals starve or

FIGURE 20–11
Hunting helps prevent wildlife from becoming overpopulated. *Courtesy of Jeff Jackson.*

die of disease. A properly controlled population is healthier and the animal population can be sustained.

Summary

In the future, agriculture will have to contend with an increasing number of people who advocate the ethical treatment of animals. This means that agriculturists will have to make sure that their management practices are proper and that they treat the animals well. In addition, they must make sure that the public is educated about how agricultural animals are treated. As with most issues, communication between the opposing sides is vital. This will help ensure that the ethically correct solution is found.

 CHAPTER REVIEW

Student Learning Activities

1. Some people feel that rodeos are cruel to animals. Research the topic and draw your own conclusion. List your reasons for making your conclusion. An alternative is to make a conclusion as to whether or not you feel that testing cosmetics on animals is justified. In class, debate others who hold the opposite opinion.

2. Make a list of all the ways animal producers protect their animals. Also make a list of management procedures that might appear to be cruel but are in the best interests of the animals. Be sure to include the reasons why they are beneficial.

3. Visit the local humane society and talk with the employees. With their help, devise a plan for your class to help stop the mistreatment of animals.

④ Talk to livestock show officials and determine what rules are in place to protect animals. Find out if there are other regulations that should be adopted.

True/False

① ___ Today in the United States, only a small percentage of the people really understand the raising of farm animals.

② ___ Most of the animals raised are produced on corporate-owned farms.

③ ___ Pigs are unsociable animals and like to be separated from other pigs.

④ ___ Most animals do better when raised with other animals of their own kind.

⑤ ___ Animals that are kept in confinement can receive better care than animals that roam free.

⑥ ___ Animals under the care and protection of humans have no need for horns.

⑦ ___ Castration keeps animals clean.

⑧ ___ Hot-iron branding is used less now than in the past.

⑨ ___ Animal rights activists believe that killing an animal is just as wrong as killing a human being.

⑩ ___ The National Pork Producers Council has adopted a Pork Producers' Creed.

Multiple Choice

① Animal welfare activists disagree with producers over what constitutes proper
    a. neutering.
    b. treatment.
    c. confinement.

② Producers are in the business of raising animals for
    a. pets.
    b. profits.
    c. insect control.

3 Animals that are placed in crates a week before they give birth are
   a. sows.
   b. cows.
   c. horses.

4 A process performed to help keep animals clean is
   a. neutering.
   b. debeaking.
   c. docking.

5 An issue of great concern for animal welfare activists is the use of animals for
   a. food.
   b. pets.
   c. research.

6 Before any drug is tested on humans, it must first be tested on
   a. insects.
   b. animals.
   c. bacteria.

7 Animals that live the best lives under the care of humans are called
   a. confined.
   b. domesticated.
   c. wild.

8 An example of a disease that has almost been eradicated through the use of research is
   a. polio.
   b. cancer.
   c. diabetes.

9 Organisms that cause stress and discomfort to animals and may even shorten their lives are called
   a. herbivores.
   b. parasites.
   c. sows.

10 A procedure that makes animals more docile and less aggressive is
   a. neutering.
   b. dehorning.
   c. confining.

Discussion

1. What are two ways of thinking concerning the well-being of animals?

2. Why is it important that animals have pleasant surroundings?

3. Why is stress so detrimental to animals?

4. How do animals benefit from confinement?

5. What management practices do animal welfare activists disagree with?

6. How have scientists justified the use of animals for research?

7. Why do scientists have a more difficult time justifying the use of animals to test cosmetic products?

8. What are the beliefs of animal rights activists?

9. Explain how deer are a good example of the human impact on controlling animal populations.

10. What is the National Cattlemen's Association statement of principles based on?

CHAPTER 21

Selecting and Using Hand Tools

Student Objectives

When you have finished studying this chapter, you should be able to:

- Practice proper safety when using hand tools.
- Explain tool selection and care.
- Identify and properly use common woodworking hand tools.

Key Terms

hand tool	ruler	boring tool
power tool	measuring tape	brace
plane	electronic measurers	hammer
customary measuring system	marking tool	screwdriver
metric measuring system	kerf	clamp
square	ripping	level
measuring device	chisel	plumb bob

Introduction

Throughout this book, you have learned about the fantastic world of agriculture. This dynamic industry represents much of the wealth of the United States. The tremendous amount of production came about as a result of research and development dealing with growing plants and animals. However, just as important as the developments in plant and animal science is the development of machinery and tools to aid in production.

As you know, agriculture began before recorded history. Thousands of years ago, humans had no tools but their bare hands. Because of their extraordinary reasoning ability, they began to devise tools that made their efforts more efficient. For example, a stone could be used as a hammer and a broken piece of flint could be used as a cutting instrument. As these crude tools were used, ideas for improvement surfaced. A stone could make a better hammer if it were shaped to fit the hand. Even more force could be applied if a stick were used for a handle. Broken flint could be shaped into a very effective knife or projectile point. Since that time, there has been an ongoing effort to make tools better.

From prehistoric time to our modern time, tools have been developed, enlarged, refined. Today, we have powerful machines, some of them two stories high, which are controlled by one person. These machines can do work that could not be done by hundreds of men without tools. Tools are multipliers of our strength, and are like millions of willing hands, working for our comfort. Without tools, there would be no civilization as we know it today.

Even through the development of complicated machine tools, **hand tools** have never lost their importance. It is still necessary in all industries that involve production and maintenance, especially agriculture. A hand tool is defined as any tool operated by hand to do work. In contrast, a **power tool**

FIGURE 21–1
In prehistoric times humans made hand tools from stone. *Courtesy of Getty Images*

FIGURE 21–2
A hand tool is any tool that is operated by hand to do work. *Courtesy of Getty Images*

is operated by some source of power other than human power. In this chapter, you will learn how to use hand tools, take care of them, and distinguish between the different kinds.

Simple Machines

From the most primitive of tools to our most complicated machine, almost all of them operate using one or a combination of six simple machines. These are the inclined **plane**, the wedge, wheel and axle, screw, lever and pulley. These simple machines are everywhere! Just look around you and you will find them everywhere.

An inclined plane is a slope that helps to move loads upward. Common examples are stair steps and ladders. Both slope upward and make movement upward easier. Another example is that of ramps that are placed at the end of a truck or trailer bed to load items such as lawn mowers or other heavy objects.

A wedge is a type of inclined plane. A wedge has a slope with a sharp edge at the end. It is usually used to separate material. An axe is a good example. The sharp edge is used to cut chips out of wood. Other types of wedges are chisels, punches and splitting wedges.

Another type of inclined plane is called a screw. Everyone is familiar with screws that are used to hold wood together. If you look closely at the screw you will see that the threads are really just a spiraling inclined plane. In fact, a screw can be considered to be a combination of an inclined plane and a wedge. As the screw is twisted, the incline plane wedges into the wood and holds it tightly. Nuts, bolts, and jar lids are other examples of screws.

A lever is a long, rigid bar used to pry. It is used to give more force such as lifting loads. A block placed under the lever is called a fulcrum. The best example of a lever and a fulcrum is the see saw on the

FIGURE 21–3
An inclined plane is a slope that helps to move loads upward. *Courtesy of Getty Images*

playground. Almost all tool handles are levers. Think about the handle in a hammer. You can hit much harder with a handle on the hammer head than you could by using the hammer head in your hand. The longer the hammer handle is, the harder the striking force. Shovels, picks, axes, hoes, and so on all make use of levers.

The wheel and axle consists of a round wheel that revolves around an axle. This simple machine is the basis for all carts, wagons, and automobiles. The principle is that the weight of an object is concentrated on the axle and the rotating movement of the wheel moves the weight. Variations of the wheel and axle are gears and pulleys. These devices are used to increase power.

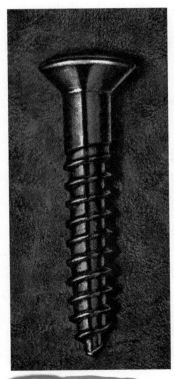

FIGURE 21–4
A screw is an example of an inclined plane. *Courtesy of Getty Images*

FIGURE 21–5
A pulley is an example of a variation of a wheel and axle. *Courtesy of Getty Images*

Working With Hand Woodworking Tools

A common beginning point to learn about tools is that of using woodworking tools. Creating projects from wood can be fun. Not only do you have the enjoyment of learning to use tools, but you can build a useful project as well. In the next sections, you will learn about some of the basic wood working tools.

Hand Tool Safety

Before we discuss the various types of hand tools, it is important that you practice the proper safety when using them. Hand tools, like any tools, can be very harmful if not used properly. Therefore, you should not just read but really learn the safety rules for any job you do and put the rules into practice.

There are certain steps that you should take when using a hand tool. They include:

- Wear safety glasses.

- Inspect tools before using. Do not use broken or damaged tools.

- Do not wear loose fitting clothing.

- Use the correct tool for the job. Do not use a tool that is either too big or too small for the job because it may result in injury or tool damage.

- Secure the work with a clamp, vise or some means of holding it. This prevents it from slipping and frees both hands to hold tools.

- Grip tools firmly.

- Stand in a safe location. You, nor anyone else, should stand directly in line with the tool's path of movement.

FIGURE 21-6
When using hand tools safety must be practiced. For example, eye protection is a requirement.
Courtesy of Getty Images

- Keep the work area clean and free of scraps and oil.
- Keep cutting tools sharp.
- Cut away from the body.
- Store tools properly. Make sure the sharp edges, if any, are down.
- Keep tools clean and free of oil and grease.

Importance of Proper Care of Tools

Knowing how to care for tools is very important, as their usefulness depends to a large degree on their care. Keeping tools well cleaned, oiled, and free from rust is essential, not only for ease in handling but also for lengthening their life. Tools should always be treated with oil when they are not in frequent use. Likewise, tools must be properly sharpened for good results. Tools that are dull and rusty will not work satisfactorily.

FIGURE 21–7
A sharp tool is much safer to operate than a dull one.

Cleaning Tools

Tools sometimes contact grease, paint, and other materials. These substances may damage tools and shorten their useful lives. You should always be careful with dirty tools, especially those that may have chemicals on them.

All grease on tools should be carefully removed with a soft material or, if necessary, by washing the tools with a safe cleaning solvent. Rust must be removed and prevented by:

1. Applying cleaning solvent and letting the tool sit for several hours.
2. Rubbing and polishing with oil and pumice stone or with an emery cloth.
3. Applying a thin film of light oil to the surface of the tool before it is put away.

FIGURE 21–8
Tools must be clean in order to operate efficiently.

Measuring and Marking Tools

There is nothing more frustrating than cutting out a piece of lumber and discovering that it doesn't fit its intended opening. The problem usually results from the worker measuring either the lumber or the opening incorrectly. To prevent this, it is important to learn how to use measuring and marking tools.

Common measuring tools are squares, rulers, and electronic measurers. The United States uses two measuring systems: the **customary measuring system** and the **metric measuring system**.

- Customary—This measuring system is used in everyday life in the United States. Inches, feet, and yards are common measurements in the customary system. A measuring device using this system is graduated, or divided, into various fractions of an inch. The common graduations used are halves, quarters, eighths, and sixteenths of an inch. The customary system is widely used in agricultural mechanics.

- Metric—The metric system is the system used by all scientists and in everyday life in most parts of the world. It is based on a decimal system, with the meter being the base for measurement. Measuring devices in the metric system are marked as meters, decimeters, centimeters, and millimeters. The metric system is not as popular in the United States as in other countries.

Before using a measuring device, check to determine the system and graduations being used. Measurements in one system can be easily converted to the other.

FIGURE 21–9
These rulers have both metric and customary scales. *Courtesy of Getty Images*

Squares

A **square** is a tool used either to check whether or not a board is square at the end or to lay out a square line. Square means that a line or the end of a board is at a perfect 90 degree angle with the rest of the board. A square may be used to lay out and check other angles as well. Always mark a board with a square before sawing it. The common kinds of squares used in agriculture are framing squares, try squares, bevel squares, combination squares, and speed squares.

- Framing square—This type of square is in the shape of the large letter "L." The longer part of the square (24") is called the body. The shorter part of the square (16") is called the tongue. Framing squares are used for squaring boards and timbers, laying out rafters, determining if the end of the board is square and squaring a board for sawing. Many different angles and patterns can be laid out using a framing square.

- Try square—The try square is smaller than the framing square and is usually from 4 to 12 inches in length. Only the blade of the try square has a measuring scale on it. The short part of the

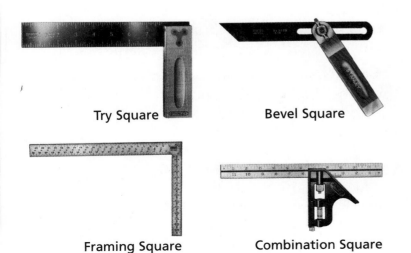

Try Square

Bevel Square

Framing Square

Combination Square

FIGURE 21–10
There are several different types of squares.

square is called the head. The primary use of this square is to determine whether or not a board is square. To protect it from being damaged and to ensure accurate measurements, the head is made of metal or wood framed in metal.

- Bevel square—These squares are adjustable so that they can be set to mark angles from 0 to 180 degrees. It is an excellent tool for marking angle cuts or checking the accuracy of angle cuts.

- Combination square—The combination square can be used for the same things as the framing

Career Development Events

If you enjoy working with tools, you will want to participate in the FFA Agricultural Mechanics Career Development Event (CDE). In this competition, you will work as a team with other students in your class. The CDE is centered on solving problems involving Agricultural Mechanics. Each of the problems in the CDE is a response to a problem that might be encountered in the workplace. According to the National FFA Organization, this CDE selects and awards those students and teams that demonstrate: (1) mastery of the subject matter and skills common to the topic; (2) effective communication skills; (3) superior problem-solving techniques: (4) an understanding of modern technology; and (5) the ability

to function as team members working together and as individuals working alone.

The National FFA Agricultural mechanics Career Development Event is divided into five areas:

1. Machinery and equipment systems
2. Industry and marketing systems
3. Energy systems
4. Structural systems
5. Environmental/natural resource systems

Individuals on each team work together and are evaluated as a team. The team is presented with a problem scenario and members are provided with the materials and equipment needed to solve the problem. Teams have to

and try squares but, because it has a movable head, its uses are more varied. It can be used as a straightedge, depth gauge, metric gauge, level, and marking gauge. Also, it can be used to lay out 45 and 90 degree angles.

● Speed square—A speed square is used for marking lumber and other materials for cutting. Information on the frame of the square often makes separate mathematical calculations unnecessary. Most speed squares have instructions for use. The use of this square is increasing in agricultural mechanics.

FIGURE 21–11
The speed square is relatively new and is helpful in laying out angles.

The Agricultural Mechanics Career Development event is centered on problem solving.

organize, assign duties, and complete all of the tasks. Individuals are also scored and given individual awards based on points scored.

This CDE, like all of the CDEs in the FFA, is designed to help you prepare job skills for the future. You learn to work as a team, communicate, and solve problems. While you are learning and developing skills, you get to travel and meet new people. At the same time, you will have loads of fun!

Measuring Devices

A **measuring device** is a tool used in making measurements. The device is normally marked in the standard units used in the measurement system. Several measuring devices are commonly used.

- Ruler—A **ruler** is a strip of wood, plastic, metal, or other material marked in increments and used for measuring and drawing lines. Rulers can be found in the customary measuring system, the metric system, or combinations of the two. The yard stick and meter stick are common examples. Most students are accustomed to using 12-inch rulers.

- Zigzag ruler—This ruler is also known as the folding ruler. The zigzag ruler is made in sections that are easily folded for convenient use. They are commonly used in making measurements on lumber in building construction.

- Measuring tape—A **measuring tape** is a ruler of metal, plastic, or other material that is wound in a case. Graduations may be in the customary or metric system. Some automatically rewind, while others are rewound with a small handle. Tape measures are the most convenient, practical, and

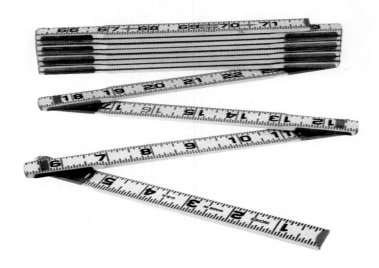

FIGURE 21–12
This is a folding or zigzag ruler.
Courtesy of Brand X Pictures

widely used measuring tool in the agricultural industry.

- Electronic measurers—**Electronic measurers,** also known as electronic tape measures, use ultrasonic sound waves to measure straight-line distances. They are accurate for a distance of up to 45 feet. Conversions are made by the device from customary to metric system and vice versa.

Marking Tools

A **marking tool** is used to make marks on wood for cutting, placing, and measuring. When marking wood, the necessary tools are pencils or knives, marking gauges, and chalk lines. The steps in marking a piece of wood are:

1. square the piece with a square.
2. mark around the entire piece.
3. saw carefully by following the line.

- Pencil or knife—Use a sharp-pointed pencil or knife to mark a board for squaring. In fine work, a very fine line is necessary because a heavy line may cause a variation in the lengths of the pieces after they are sawed. This differ-

FIGURE 21–13
Tape measures are convenient and easy to use. *Courtesy of Thinkstock*

FIGURE 21–14
Pencils are used to mark wood at the proper place. *Courtesy of Getty Images*

ence may be important when fitting small pieces together.

- Marking gauge—A marking gauge is often used when marking lines parallel to the edges of the material. However, this job may also be done with a straightedge or a pencil.

- Chalk line—A chalk line is used for marking a straight line on a board, wall, ceiling, or floor. It is a string that has been coated with colored chalk dust so that it leaves a line when snapped.

Saws

Handsaws are the most common cutting tools for wood used by agricultural workers. The handsaw is used to cut larger pieces of lumber into smaller pieces, according to the needs of specific jobs. The most common types used in agriculture are crosscut saws, ripsaws, compass saws, and coping saws.

Saws are generally categorized according to the number of points or teeth they have per inch of blade. An 8-point saw, for example has 8 teeth for every inch of blade, whereas a 10-point saw has 10 teeth for every inch. The more points per inch, the finer or smoother the cut made by the saw.

A cut made by a saw is called a **kerf**. To start the kerf with the saw, it is important for the blade to be on the side of the line that will become the scrap piece of lumber. Then the saw is drawn backward in a smooth stroke. Long, smooth forward and backward strokes are then made, following the line drawn. The blade must always be on the side of the waste lumber. The last few strokes are made very slowly and with slightly less pressure. This helps prevent the board from splitting as the cut is completed.

- Crosscut saws—As the name indicates, these saws are used for cutting across the wood grain. A 10-point crosscut saw is the most commonly used in the agricultural industry. However,

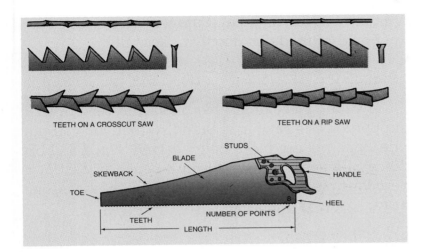

TEETH ON A CROSSCUT SAW

TEETH ON A RIP SAW

STUDS
BLADE
SKEWBACK
HANDLE
TOE
HEEL
TEETH
NUMBER OF POINTS
LENGTH

FIGURE 21–15
Hand saws are used for cross cutting or ripping, depending on the shape and angle of the teeth.

crosscut saws are also available in 9 and 12 points. The crosscut saw is held at a 45 degree angle to the wood being cut, and the other hand is used to steady the material.

- Rip saws— **Ripping** a board means to cut it lengthwise along the grain to make strips of narrower width. The rip saw is used for this work. Its teeth are coarser, with 5½ points per inch being the most common. The teeth chisel the wood rather than cut it as the crosscut saw does. Because of this difference, the saws are sharpened differently. The ripsaw works best if held at a 60 degree angle to the material being cut.

- Compass saw—These saws are used for cutting curves, circles, and irregular shapes in pieces of lumber. Before a circle or other shape can be cut in wood, it is necessary to drill a hole large enough for the end of the blade to penetrate and start the kerf.

- Coping saw—This saw is also used for cutting curves or making other irregular cuts. The coping saw is different than the compass saw in that its blade is thin and removable and it has a great many more teeth along the blade so that it can

FIGURE 21–16
A coping saw is used for cutting curves or other irregular cuts.
Courtesy of Getty Images

make a much finer cut. Cuts with the coping saw should be made on down strokes only.

Holding Lumber in Place for Sawing

Place a long piece of lumber on sawhorses or other supports while sawing it. Stand so that you can put one knee on the piece being sawed. Place a short piece of wood in a vice or on a bench hook. For a very long piece of lumber, use two bench hooks, one at each end of the piece.

Wood Chisels

A **chisel** is a wedge-shaped cutting tool. It is designed to make special cuts in wood, such as grooves or notches. Chisels are also used to shape and trim wood.

The chisel consists of a blade and handle. The cutting edge of the blade is beveled, or angled. The opposite end of the blade is called the tang. This end fits into the handle. The two types of chisels frequently used in carpentry work are tang chisels and socket chisels.

The tang chisel is one in which the tang extends into a wooden or plastic handle. In a socket chisel,

FIGURE 21–17
A wood chisel is used to trim and shape wood.

the handle is fitted into a socket formed by the metal in the chisel.

Chisels are used by workers by pushing the chisel forward, with the bevel (angle) down and the chisel positioned at an angle to the wood. The chisel can also be driven forward by tapping the handle or socket with a mallet or hammer. The degree of the angle usually determines how much wood is to be chiseled in one forward motion. If the chisel is held at a high angle, more wood will be cut away.

Boring Tools

A **boring tool** is a tool used to make holes or change the size or shape of holes. Boring tools include bits, drills, reams, and the devices used to turn them. The most common boring tool is the brace and auger bit.

A **brace** is the device for holding and turning an auger bit. Braces can either be the ordinary type or the ratchet type. An ordinary brace is satisfactory for all general work where there is plenty of room for making complete revolutions with the handle. A ratchet brace is constructed so that complete revolutions of the handle are not necessary. Consequently, it is convenient for use in corners or other places with little room to work.

The hand drill and electric drill are other tools used for boring holes in wood. Bits for the electric drill operate at high speeds and are different from bits used in the hand drill and the brace.

Hammers

A **hammer** is a tool made of durable metal with a handle made of wood, metal, or other material. There are two kinds of nailing hammers: one has a curved claw and the other has a straight claw. They are often referred to as a curved-claw and ripping-claw hammer. The curved claw is particularly valuable for

FIGURE 21–18
A brace and bit is used to bore holes in wood. *Courtesy of Getty Images*

[Handwritten margin note:] Hammers are classified by: ① Type of Claw ② Striking Face ③ Handle.

pulling out nails. The ripping claw is best used to rip apart or tear down wooden structures and to pry two boards apart.

Hammers have two kinds of striking faces. One is flat, and the other is slightly rounded. The slightly rounded head, called a bell-faced hammer, is used primarily to drive a nail slightly below the surface of the wood without leaving hammer marks.

Hammers are also classified by the kind of handle. Hammers with wooden handles are lighter and used in framing and general carpentry. Those with steel handles are used for heavier work. Hammers with steel handles and straight claws are used for

Supervised Agricultural Experience

If you have an interest in tools and hardware, you could benefit from completing an SAEP at a hardware store. Supervised Agricultural Experience Programs (SAEPs) are designed to provide hands-on experience for students to develop their skills in agricultural career areas. To begin, you must first be enrolled in an agriculture class. Your teacher will help you to design a SAEP that combines your interests and skills learned in the classroom, shop, and laboratory. By becoming active in SAEP, you will develop skills in your area and learn how to use those skills on the job. You will also learn how to earn, save, and invest money.

A SAEP in a hardware store is a type of placement program. Placement programs involve placing students on farms, in agricultural businesses, in school labs, or in community facilities to provide a "learning by doing" environment. This experience is done outside of normal classroom hours and may be paid or nonpaid. A placement SAE in a hardware store requires supervision by agriculture education teachers, employers, and in cooperation with parents. These people work with the student and assist them in their development and achievement of their educational and career goals.

Along with the responsibility of maintaining a SAE comes recordkeeping.

ripping. Hammers with fiberglass handles are used for general work, depending on claw shape.

Screwdrivers

A **screwdriver** is a tool with a handle on a long metal shank. Screwdrivers come in all shapes and sizes and are designed to fit the heads of a variety of screws. Whenever materials are to be joined together with screws, and that happens in nearly all agricultural career fields, the screwdriver is needed to drive the screws.

There are basically two kinds of screwdrivers for woodworking, the standard and the Phillips. The

FIGURE 21–19
A ripping claw hammer (above) is used to rip apart nailed boards. A curved claw hammer is used to pull nails. Both are also used to drive nails. *Courtesy of Brand X Pictures # brxbp39393 and Getty Images*

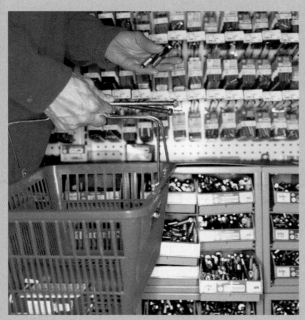

A SAEP in a hardware store can help you learn a lot about tools.

Recordkeeping is one of the most important components of a SAEP whereupon students maintain a list of skills learned during the time spent in the classroom and at work. Material inventories, income, expenses, and accomplishments should be kept along with managerial decisions, labor, and advancements made on the job. With a long-term SAEP, good recordkeeping, and FFA involvement, a student can earn well-deserved recognition. Proficiency awards for superior SAEPs are among the many honors earned through agricultural education.

FIGURE 21–20
The screwdriver on the left is a Phillips and the one on the right is a standard screwdriver. *Courtesy of Getty Images*

standard head is flat and tapered. The Phillips screwdriver has a cross-shaped bit.

The screwdriver depends on the head of the screws to be driven. The tip of the screwdriver should not be wider than the screw. If it is wider, it will scar the wood around the head of the screw. The tip of the screwdriver should not be larger or smaller than the slot in the screw. If it is larger, it will not extend to the bottom of the slot in the screw; if it is smaller, it will slip out of the slot easily.

No tool is perhaps abused more than the screwdriver. It should not be used other than to drive screws. Excessive force exerted on the long handle of a screwdriver can snap the handle. This could in turn cause hand or bodily injury.

Clamps

A **clamp** is a device for holding pieces of wood or metal in position until fastened together by nails, screws, bolts, glue, or welding. Clamps may also be used to exert pressure so that materials may be fitted together properly. Common types of clamps are hand screw clamps, "C" clamps, bar clamps, and miter clamps.

Levels

A **level** is used to determine whether a surface is flat (horizontal). Levels involve using a bubble in a freeze-resistant liquid in a small tube-shaped, glass container. Markings on the tube are used to assess the position of the bubble and determine if the surface is level. The bubble is entirely between the two markings if the surface is level.

Levels are in various lengths, ranging from a few inches to over four feet. They are made of wood or metal. Metal levels are often preferred because they

FIGURE 21–21
Clamps are used to hold wood together for gluing or nailing. *Courtesy of Getty Images*

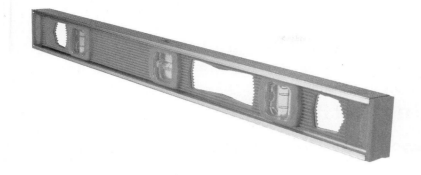

FIGURE 21–22
A spirit level uses a bubble enclosed in glass to indicate when an object is level or plumb. *Courtesy of Brand X Pictures #brxbxp39439 and Digital Vision dv1450016*

do not warp. In selecting a level, choose the longest one that is convenient for the job. A longer level is more accurate in determining the surface position.

Plumb Bob

A **plumb bob,** also known as a plummet, is used to determine a plumb, or vertical, line. A plumb line is at a right angle to a level horizontal line. This is important in building construction to assure that the walls, floors, and ceilings fit properly. A plumb bob may be used to find a point directly beneath an overhead point. A breeze may destroy the accuracy when a plumb bob is being used by moving it around.

Summary

Humans have used hand tools since prehistoric times. Tools help us to work more efficiently and also to do work we could not do with our bare hands. Today there is a wide variety of different hand tools to choose from. Learning to identify and properly use hand tools will allow us to build, maintain, and repair our homes in an effective way. Safety should always be practiced when using any type of tool. Tools used safely and properly will give many years of service.

FIGURE 21–23
A plumb bob is used indicate a vertical line. *Courtesy of Getty Images*

CHAPTER REVIEW

Student Learning Activities

1. Closely examine a bicycle. Make a list of all the simple machines that make up the bicycle.

2. Use the Internet to explore kinds and sources of hand tools. Several Web sites are listed below. The Hand Tool Mall has excellent safety information.

   Matco Tools—http://www.matcotools.com

   The Stanley Works—http://www.stanleyworks.com

   Hand Tool Mall of the Internet—http://www.toolsource.com

   Hand Tools Institute—http://www.hti.org

   Old Forge Tools—http://www.oldforge.com

3. Prepare a bulletin board with pictures of the various hand tools described in this chapter and captions with descriptions of their uses.

True/False

1. ___ You should wear loose fitting clothing when working with hand tools.

2. ___ Tools that are cleaned, oiled, and free from rust aid in ease of handling.

3. ___ Tools play a small role in agriculture.

4. ___ The problems resulting in measuring and marking are usually the result of the person who did the measuring.

5. ___ All handsaws are made the same.

6. ___ The cutting edge of a chisel is an example of a wedge.

7. ___ Braces are used to bore holes in wood.

8. ___ Two types of hammers are curved claw and straight claw.

9 ___ A hammer is needed to drive screws in agriculture.

10 ___ A plumb bob is also known as a plummet.

Multiple Choice

1 A weighted string used to find a level vertical line is called a:
   a. level
   b. plumb bob
   c. square

2 Levels make use of a glass tube filled with:
   a. water
   b. salt
   c. a freeze resistant liquid

3 Standard and Phillips are types of:
   a. screwdrivers
   b. saws
   c. hammers

4 Hammer handles can be made of:
   a. wood
   b. fiberglass
   c. wood or fiberglass

5 A device used to turn an auger bit is called a:
   a. brace
   b. clamp
   c. twister

6 Ripsaws should be held at an angle of:
   a. 40 degrees
   b. 80 degrees
   c. 60 degrees

7 Electronic measurers use _____ to measure straight line distances.
   a. electronic sound waves
   b. microwaves
   c. radio waves

8 A square that is used for laying out rafters is called a:
    a. try square
    b. combination square
    c. framing square

9 Square refers to an angle of:
    a. 90 degrees
    b. 45 degrees
    c. 60 degrees

10 A measuring system that uses meters, centimeters, and millimeters is called the:
    a. standard system
    b. customary system
    c. metric system

Discussion

1 Why is the proper care of tools important?

2 Name the different types of simple machines.

3 What are the common kinds of squares?

4 What is the difference between a rip saw and a crosscut saw?

5 What is a kerf?

6 What tools are commonly used for marking?

7 What is a level used for?

8 What tool is abused the most frequently?

9 How can you classify hammers?

10 What are plumb bobs used for?

CHAPTER 22

Small Engine Operation

Student Objectives

When you have finished studying this chapter, you should be able to:

- Discuss the uses of small engines.
- Observe all safety precautions when working with small engines.
- Distinguish between two- and four-cycle engines.
- Explain how a four-cycle engine operates.
- Explain how a two-cycle engine operates.
- Identify all the systems in a small engine.
- Explain how the different systems of a small engine work.

Key Terms

cycle	internal combustion	power stroke
stroke	engine	compression stroke
four-stroke cycle engine	engine timing	intake stroke
two-stroke cycle engine	top dead center (TDC)	scavenging
engine	bottom dead center (BDC)	magneto ignition system
motor	compression ratio	
combustion	exhaust stroke	

Perhaps the greatest impact on agricultural production was the invention of the **internal combustion engine.** This type of engine burns fuel inside the engine. Until its invention, engines burned fuel outside the engines. These engines usually produced steam for a power source. We generally think of internal combustion engines as running large tractors and harvesting equipment. However, a huge number of small engines are in use. These are usually one cylinder engines that are portable and run small equipment.

Today, there are over 80 million small air-cooled engines in use in the United States. These small engines are a popular power source for many different types of equipment.

One reason why small engines are so widely used is because the engines are easily adaptable to many jobs and conditions. They are easily adaptable because:

1. The engines require no outside source of power, therefore, they are compact and relatively lightweight.

FIGURE 22–1
There are over 80 million small engines in use in the United States.

2. They are cooled by air instead of liquid coolant.
3. The engines are relatively easy to service and repair, which allows operators to do their own maintenance.

Examples of equipment powered by small engines are: lawn mowers, chainsaws, go-carts, mini-bikes, tillers, pumps, and sprayers.

Small Engine Safety

Before the operation and mechanics of small engines is discussed, it is important to address safety. Annually in the United States, 150,000 people are injured while mowing grass. Over 55,000 are treated in hospitals. There are 55,000 toes and 18,700 fingers amputated by lawnmower blades. Only 9 percent of these injuries were caused by mechanical failure, whereas 91 percent were caused by human error.

Some of the common ways people get hurt by small engines include touching a power mower's swirling blades with a hand or foot, being struck by a flying object hurled by rotary mower blades, catching a body part while operating or repairing a machine while the power is on, having a machine overturn, falling off a machine, and receiving an electrical shock.

In order to prevent minor or severe injuries while operating small engines, it is important that certain safety precautions be taken. Operators should adhere to the following accident prevention reminders:

1. Dress for the job. Wear safety glasses and leather shoes.
2. Keep hands and feet a safe distance from all moving parts.
3. Do not allow burning material to get near oil, solvents, or gasoline.

FIGURE 22–2
Saftey should be foremost in operating small engines. Never fill a fuel tank when the engine is hot.
Courtesy of Thinkstock.

4. Stop the engine and allow it to cool before refueling.
5. Whenever possible, handle gasoline outdoors.
6. Disconnect the spark plug cable when making any adjustments on machines to prevent the possibility of accidental engine startup.
7. Never disengage any of the safety equipment or devices.
8. Keep all shields in place.
9. Do not operate engines above the specified speed.
10. Do not overload engines or force equipment beyond the designated capacity.
11. Never start and run an engine inside a building in which the ventilation is inadequate.
12. Never use gasoline as a solvent or to clean engine parts.
13. Dispose of gasoline properly by pouring it into an approved container. Never pour it onto the ground or down a drain.

Importance of Proper Care of Small Engines

Small engines are rugged machines that are built to last. Most small engines are designed to operate continuously at or near top operating speeds to prevent overload. Small engines are designed and built to operate under full load at top operating speeds for at least 1,000 hours without requiring major repairs, if the engine is properly serviced and maintained. Modern small engines take little time and effort to maintain.

The most important aspects are keeping the oil at the correct level and keeping the air breather clean. Before you operate or attempt maintenance on a small engine, read and thoroughly understand the owner's manual.

FIGURE 22–3
Modern small engines take little time and effort to maintain.
Courtesy of Getty Images

Operating Principles of Small Engines

A person who intends to service or repair a small engine must have at least a working knowledge of the operating principles of the engine. Before you can trouble shoot problems or fix the engine, you have to understand how the engine operates. You need to understand the operation of the various parts of the engine and be able to identify each step in the operation cycle. Once you know this, locating and fixing problems will be much easier.

Power Production

An **engine** is a device or machine that can produce power on its own—independent from an external source of energy. An engine has an energy source (fuel) and converts the fuel to useable power. The power produced is usually in the form of mechanical power.

How is an engine different from a motor? Generally, a **motor** is considered to be a device or machine that produces power from the energy supplied by an external source. For example, an electric motor takes the electrical power produced somewhere else, such as a generator, and then converts it to mechanical power.

Combustion

Combustion is the chemical process of oxidation that produces heat. The combustion process requires three things: fuel, air (oxygen), and heat. An internal combustion engine uses the process of rapid combustion to operate. Sufficient heat must be produced from the combustion process to produce useable power.

FIGURE 22–4
An engine produces power using energy generated within the device. A small engine is a good example.

FIGURE 22–5
A motor produces power using an outside source. An electric motor is an example.

Internal Combustion Engines

Internal means that the combustion occurs in an enclosed chamber (called the combustion chamber). The heat produced is converted to mechanical power. An internal combustion engine is a machine or device that is capable of converting heat energy into mechanical power. The mechanical power produced is usually in the form of a rotating force.

The small engine, like any gasoline burning engine, is an internal combustion engine. Four events must take place within the engine for an internal combustion engine to run continuously.

Career Development Events

One of the parts of a high school Agricultural Education program is Supervised Agricultural Experience (SAE). This part of the program is designed so that students can put into practice what they learn in the classroom. These activities are planned, practical activities conducted outside the classroom. They may be conducted at the student's home or may be organized at a business such as a small engine repair shop. The experience is called *supervised* for a good reason. The agriculture teacher visits with the student at the site of the SAE to give advice and help direct the experience. SAE programs are explained in more detail in Chapter 25.

The National FFA Organization provides awards for students who excel in their SAE. Each award is called a Proficiency Award, and they all are based on the goals achieved by the student and the skills and concepts learned. They are also based on the records kept by the student. The Agricultural Mechanics Design and Fabrication proficiency is one of three agricultural mechanics proficiency awards available to FFA members. The others include "Agricultural Mechanics Repair and Maintenance" and "Agricultural Mechanics Energy Systems." These proficiency awards are presented to individual FFA members who have effectively produced and marketed their talent in

The four events and the order in which they occur are:

1. Intake—the process of getting the fuel and air required for combustion into the combustion chamber.
2. Compression—the process of compressing the fuel and air mixture in the combustion chamber.
3. Power—the ignition, burning, and expansion of the fuel-air mixture.
4. Exhaust—removing spent products of combustion from the combustion chamber.

A Proficiency Award in Agricultural Mechanics rewards students who excel in their Supervised Agricultural Experience Program (SAEP).

the agricultural mechanics industry. The student must have a Supervised Agricultural Experience (SAE) in a selected special agricultural mechanics area. All of the agricultural mechanics Proficiency Award areas are in the entrepreneurship/placement category. This means that the student must be the entrepreneur of his/her own agricultural mechanics project or work in the agricultural mechanics production area.

Planning and goal-setting for winning proficiency awards should begin as soon as a student enters the Agricultural Education program. The SAE can last until the student has been out of high school for a year. Talk with your teacher and begin planning your SAE and plans for winning a Proficiency Award.

These four events make up an internal combustion engine cycle. They must occur in the proper sequence and at the proper time. **Engine timing** refers to the proper timing and sequence of these events. Once a cycle is completed, another cycle immediately begins.

Terms Used to Describe the Operation of Engines

There are four terms used to describe the operation of small engines. They include:

1. **Cycle** —the completion of the four events: intake, compression, power, exhaust.
2. **Stroke** —the full travel of the piston in one direction, either toward the crankshaft or away from the crankshaft.
3. **Four-stroke cycle engine** (four-cycle engine)— an engine requiring four strokes of the piston to complete one cycle. One stroke of the piston is required for each of the four events.
4. **Two-stroke cycle engine** (two-cycle engine)— an engine requiring only two strokes of the piston to complete a cycle of operation.

The Four-Stroke Cycle Engine

The four-stroke cycle engine operates on a series of four strokes, or piston movements, to a cycle. The piston moves up and down within a cylinder that is only slightly larger than the piston diameter.

The cylinder is inside a mass of metal (usually aluminum) called a block. At the top of the block are valves that let in fuel and let out exhaust fumes. The piston is connected to the crankshaft that converts the piston's up-and-down motion to a rotary motion, much like a person pedaling a bicycle. The

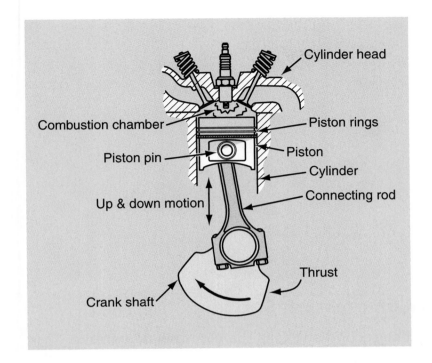

Combustion chamber

Piston pin

Up & down motion

Crank shaft

Cylinder head

Piston rings

Piston

Cylinder

Connecting rod

Thrust

FIGURE 22–6
The main parts of a 4-cycle engine.
Note that the valves on this one
are in the head. Some small engine
valves are in the block.

up-and-down movement of the person's legs is converted to a rotary movement of the bicycle wheels.

As the pistons move up and down, the movement in one direction is called a stroke. When the piston reaches the top of its stroke, the volume of the combustion chamber is at its smallest. This position is called **top dead center** (TDC). When the piston reaches the bottom of the stroke, the cylinder volume is at its greatest. This position is called **bottom dead center** (BDC). Each of these strokes serves a particular purpose.

Intake Stroke

The first stroke of the cycle is the **intake stroke.** This occurs when the piston moves down, the intake valve opens, and a fuel mixture of fuel and air enters the cylinder. At the completion of this stroke, the piston is at the bottom of the cylinder and both valves are closed.

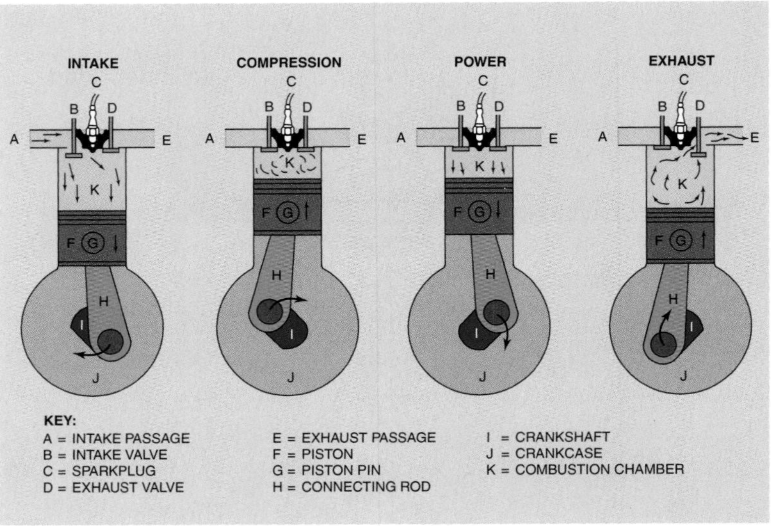
INTAKE
COMPRESSION
POWER
EXHAUST
A
B
C
D
E
F
G
H
I
J
K
KEY:
A = INTAKE PASSAGE
B = INTAKE VALVE
C = SPARKPLUG
D = EXHAUST VALVE
E = EXHAUST PASSAGE
F = PISTON
G = PISTON PIN
H = CONNECTING ROD
I = CRANKSHAFT
J = CRANKCASE
K = COMBUSTION CHAMBER

FIGURE 22–7
This diagram shows all of the strokes of a four cycle engine. A complete 4 stroke 2 cycle is completed 25 times every second

Compression Stroke

The second stroke is the **compression stroke.** This occurs when the piston reaches the bottom of the cylinder on the intake stroke and starts upward. Gasoline engines compress an air-fuel mixture. At the top of this stroke, the air-fuel mixture has been compressed into the area in the head called the combustion chamber. The degree that the air-fuel mix is compressed is referred to as the **compression ratio.** This is calculated by comparing the volume of the cylinder at TDC to the volume of the cylinder at BDC. The volume of a cylinder is calculated by the formula $V = \pi r^2 h$, where $\pi = 3.14$, r = the radius of the cylinder, and h = the height of the cylinder. If the diameter of the piston is 2½ inches, the radius would be 1¼ inches (radius is one-half the diameter). If the height of the cylinder at TDC is .5 inches, the volume at TDC would be 1.53 cubic inches. If the height of the cylinder at BDC is 4 inches, the volume at BDC would be 12.25. The volume of the cylinder at BDC (12.25) divided by the volume at TDC (1.53) is 8.00, or an 8:1 compression ratio.

Power Stroke

As the piston nears the top of the cylinder on the compression stroke, a spark from the spark plug ignites the air-fuel mixture. The spark is precisely timed by the ignition system to occur at exactly the correct time. The resulting rapid expansion forces the piston down, creating the **power stroke.** During the power stroke, the intake and exhaust valves must remain tightly sealed. The larger the cylinder and piston, the more power the engine can produce. This is because there is more room for the air-fuel mixture and, the larger the room, the greater the amount of energy that is released on ignition of the gases.

Exhaust Stroke

The fourth and final stroke of the cycle is the **exhaust stroke.** This occurs when the piston reaches the bottom and starts upward. The purpose of the exhaust stroke is to remove the burned air-fuel mixture so fresh air and fuel can be brought into the combustion chamber. During this stroke, the exhaust valve opens and, as the piston moves up, the exhaust gases are forced out through the exhaust system. The process of moving the exhaust gases is called **scavenging.** For an engine to operate efficiently, the exhaust must be scavenged as completely as possible.

At the end of the exhaust stroke, the piston is at TDC, and a new engine cycle begins. It is important to keep in mind that a complete cycle of the four-stroke cycle engine happens in a very short period of time. For example, if a single cylinder engine is operating at a speed of 3,000 revolutions per minute (rpm), the engine is operating at 50 revolutions per second. Therefore, the engine is completing 25 cycles per second and each piston stroke is taking $\frac{1}{100}$ of a second to complete.

FIGURE 22–8
Two-cycle engines are light-weight and can be operated in any position. Note the safety precaution of the operator. She is wearing hearing protection, eye protection, and is safely dressed.

The Two-Stroke Cycle Engine

The two-stroke cycle engine is also widely used in the agricultural industry. This type of engine completes the intake, compression, power, and exhaust stages in two strokes. It is also a piston- or reciprocating-type internal combustion engine. Generally, the operation of the two-stroke cycle engine is more difficult to understand, because more than one engine cycle is occurring at the same time. Two-stroke cycle engines are commonly used in agriculture for small engine applications because they are more lightweight and can be operated upside down. Chain saws, weed eaters, leaf blowers, and other small power tools are examples of implements that use two-cycle engines.

The two-stroke cycle engine operates on the principle of using the movement of both ends of the piston to create compression and a partial vacuum. The top of the piston is used to operate and create the combustion chamber in the cylinder. The bottom of the piston is used to draw the air-fuel mixture into the engine through an opening in the cylinder wall. This opening is called a port.

Intake and Exhaust Strokes

As the piston nears the bottom of its stroke, it uncovers the intake and exhaust port(s). Because the air-fuel mixture in the crankcase is under pressure, it rushes through a passage to the intake port and enters the cylinder. This incoming gas mixture pushes air, or exhaust, out of the cylinder. Therefore, intake and exhaust functions occur with very little movement of the piston.

Compression Stroke

At the end of the intake/exhaust stroke, the cylinder is filled with an air-fuel mixture. The piston moves

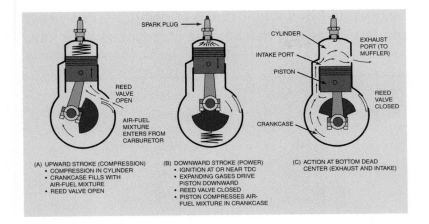

FIGURE 22–9
The strokes of the two-cycle engine are illustrated here.

upward, closes the intake and exhausts ports, and compresses the air-fuel mixture trapped in the cylinder. At the same time, a new supply of air and fuel rushes into the crankcase.

Power Stroke

The spark plug fires to ignite the mixture at or near TDC. The burning and expanding gases drive the piston downward through the power stroke. This same downward movement puts pressure on the new air-fuel mixture in the crankcase. Thus, the engine completes its cycle of intake, compression, power, and exhaust with only two strokes of the piston.

Comparing Four- and Two-Stroke Cycle Engines

The four-stroke and two-stroke cycle internal combustion engines have specific characteristics that allow both to be used for various agricultural purposes.

Generally, the four-stroke cycle engine operates more quietly, is heavier, has a longer life, and is cleaner burning than a two-stroke cycle engine. However, the two-stroke cycle engine is lighter, can be operated in a wider range of positions, has fewer moving parts, and can be used more for smaller jobs.

Two-Stroke Cycle Engines	Four-Stroke Cycle Engines
Lighter weight	Heavier weight
Operates in many positions	Operates in limited positions
Higher power-to-weight ratio	Lower power-to-weight ratio
Engine oil usually mixed with fuel	Engine oil in a reservoir
Louder operation	Quieter operation
Higher engine speeds	Slower engine speeds
More vibration	Smoother operation
Rough idling operation	Smoother idling operation

Diesel Engines

Many small engines are diesel engines. In the past, diesel engines have been mostly used for large equipment and trucks. However, in recent years, diesel engines are made that are smaller and lighter than the older heavy diesels. Small diesel engines are used to operate little tractors, generators, and irrigation pumps.

These engines differ from gasoline engines in that they have a different ignition system. Recall that the fuel mixture in a gasoline engine is ignited by a carefully timed spark from a spark plug. Diesel engines are ignited by the heat of compression. When air is tightly compressed, it becomes hot. Rapidly pump a bicycle pump and feel the bottom of the pump. The heat you feel is the heat of compression. In a diesel engine, an injector is used instead of a spark plug. A high pressure stream of diesel fuel is injected into the cylinder at the end of the compression stroke. The heated air causes the fuel to ignite and the power stroke begins. The compression ratio for a diesel is much higher than that of a gasoline engine in order to get enough heat to ignite the fuel.

The advantages of diesel engines are that they are usually tougher, produce more power, and last longer than gasoline engines. In addition, fuel costs are generally lower. The disadvantages of diesel engines are that they are heavier, they are noisy, and they often have an unpleasant odor. Despite these disadvantages, small diesel engines are becoming more popular.

Engine Systems

An engine operates through the coordination of several different systems that must function properly for the engine to operate. These systems are much like the systems of the body. Our digestive, skeletal, circulatory, and nervous systems must all function properly before the body can operate efficiently. All systems of the body rarely get sick at the same time. Likewise, when an engine malfunctions, it is usually only a part of one system that malfunctions. That part, however, can cause the entire engine to malfunction.

The Compression System

The primary purpose of this system is to compress the air to increase the energy resulting from the combustion of the fuel. The compression system of the internal combustion engine uses pistons that move up and down in a cylinder. The piston and cylinder must form a leakproof combustion chamber for the engine to operate. The air tightness of

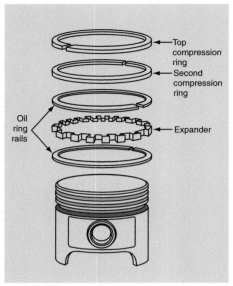

FIGURE 22–10
Rings set in grooves on the piston expand against the cylinder wall to create a tight fit.

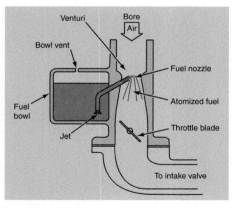

FIGURE 22–11
Most small engines have a device called a carburetor that mixes gasoline with air in the proper ratio.

the combustion chamber is a major design feature for internal combustion engines. If the compression system loses compression, the engine will not operate properly. To ensure that the fit is tight, the pistons have grooves all the way around the top of the piston. Inside these grooves are rings of steel or cast iron that expand against the cylinder wall and create a tight fit.

The Intake System

The purpose of the air intake system is to allow clean air to efficiently enter the combustion chamber. Air is composed of approximately 20 to 25 percent oxygen. Oxygen is required for combustion. The air intake system cleans dirt particles from the air without causing a significant restriction to the free flow of the air, which would decrease engine power output.

Fuel Systems

The fuel system is designed to deliver clean fuel to the combustion chamber and only allow the amount of fuel necessary for efficient operation. The fuel needs to be sent at the correct time and in the best proportion with the air for the engine to operate properly. There are a variety of fuel systems and components to meet the needs of different kinds of engines, fuels, and applications. However, most fuel systems operate based on the same basic principles. Small engines have a device called a carburetor that mixes the gasoline with air.

Suction from the intake stroke sends the fuel mixture into the combustion chamber.

Exhaust Systems

The exhaust system performs the task of removing the burned gases from the combustion chamber. The exhaust system also helps with engine noise reduction and heat transfer. A device called a muffler is attached to the cylinder. The exhaust fumes pass through the muffler. This process "muffles" or deadens a lot of the sound from the engine. In recent years, the exhaust system has been designed to help control engine emissions that may damage the environment.

Cooling Systems

Engine cooling systems are designed to manage the heat produced by the combustion of the air and fuel. Every engine has an operating temperature that is best for that engine. The task of the cooling system is to allow the engine to reach the correct temperature and then maintain this temperature under varying conditions.

Almost all small engines are cooled by the air. Most large engines are cooled by a system that circulates liquid coolant through the engine. In air cooled systems, the heat from the engine components is transferred directly to the surrounding air. This process is aided by fins that are made into the cylinder black. The fins provide a larger surface area that comes in contact with the cooling air.

FIGURE 22–12
Fins on the engine block and head help dissipate heat from the engine.

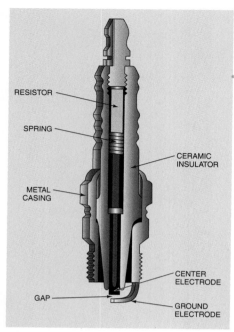

RESISTOR

SPRING

METAL
CASING

GAP

CERAMIC
INSULATOR

CENTER
ELECTRODE

GROUND
ELECTRODE

FIGURE 22–13
Current flowing through the spark plug tries to jump across the gap at the end of the spark plug. This creates a spark.

Ignition Systems

Ignition systems are designed to ignite the air and fuel mixture in the combustion chamber. Spark ignition engines (gasoline) use an electrical spark to ignite the air and fuel mixture. These systems create an electrical spark in the combustion chamber at the end of the compression stroke. With correct ignition timing, the ignition spark occurs at the precise moment to insure the most complete and efficient combustion of the air-fuel mixture. The electrical spark is generated from a magnetic field created by magnets. This system is called a **magneto ignition system.** A device called a coil increases the amount of voltage (electrical pressure) generated by the magneto. The current flowing through the spark plug can be around 40,000 volts.

This current flows through the spark plug and tries to jump across the gap at the end of the spark plug. This creates a spark.

Lubrication System

Internal combustion engines need lubricants to keep engine parts operating smoothly. The engine uses a lubrication system to keep the engine parts lubricated to reduce friction, remove heat, keep engine parts clean and provide tight seals on various engine parts to keep dirt out and compression pressure in. The lubrication system is designed to maintain the optimum amount of engine oil lubricant on various parts. Even small engines have lubrication systems, although they may not be as complex as the systems used on larger engines. It is extremely important to keep the engine filled to the proper level with clean oil. This means that the oil must be checked before the engine is started. Oil should be changed at intervals suggested by the manufacturer of the engine.

Starting Systems

All internal combustion engines require some type of system to start. Some small engines use a manual operation system, such as a recoil rope starter, which the operator pulls; an example of this type of system is a go-cart. Most large engines use an electrical starting system, which uses an electrical motor to start the engine.

Summary

Small engines are an important part of the agricultural industry. Although they exist in a variety of types, they all operate basically the same way. They may seem complicated, but they are made up of several less-complicated systems. An understanding of the basic principles of operation and how the systems work together is essential to the operation, maintenance, and repair of the engines.

22 CHAPTER REVIEW

Student Learning Activities

1. Disassemble a small engine and label the parts. Reassemble the engine.

2. Create a list of the various small engines on a farm, school facility or agribusiness. Determine whether the engines are two- or four-stroke engines and whether they are gasoline or diesel engines.

True/False

1. ___ Small engines require an outside source of power.

2. ___ Nine percent of small engine injuries are caused by the operator.

③ ___ An engine can produce power on its own.

④ ___ Internal combustion means that combustion occurs in an enclosed chamber.

⑤ ___ The four terms used to describe the operation of small engines are cycle, stroke, combustion and four-stroke cycle engine.

⑥ ___ Two-stroke cycle engines are a lighter weight.

⑦ ___ The exhaust stroke of a four-stroke cycle engine occurs when the piston reaches the bottom and starts upward.

⑧ ___ The two-stroke cycle diesel engine is common in agriculture.

⑨ ___ Lubricants are used to keep engine parts running smoothly.

⑩ ___ Most large engines use an electrical starting system.

Multiple Choice

① How many hours can small engines operate without requiring major repairs?
    a. 500
    b. 1000
    c. 1500
    d. 2000

② What does the combustion process NOT require?
    a. fire
    b. fuel
    c. air
    d. oxygen

③ _____ is the full travel of the piston in one direction, either toward or away from the crankshaft.
    a. cycle
    b. intake
    c. stroke
    d. rotation

4 When the piston reaches the top of its stroke, it is said to be at _____
   a. the top
   b. dead center
   c. top dead center
   d. bottom dead center

5 What is the purpose of the exhaust stroke?
   a. remove the unwanted gas
   b. remove the fumes in the cylinder
   c. discard the excess air
   d. removed the burned air-fuel mixture

6 Which of the following is an example of a piece of equipment that operates using a two-stroke cycle engine?
   a. lawnmower
   b. go-cart
   c. leaf blower
   d. mini-bike

7 The purpose of the _____ system is to allow clean air to enter the combustion chamber.
   a. intake
   b. fuel
   c. compression
   d. ignition

8 What does a lubrication system NOT do?
   a. remove heat
   b. reduce friction
   c. reduce speed
   d. keep parts clean

9 Most large engines use a(n) _____ starting system.
   a. gasoline
   b. electrical
   c. spark ignition
   d. key-driven

⑩ Annually, in the United States, _____ people are injured mowing grass.
   a. 50,000
   b. 100,000
   c. 200,000
   d. 150,000

Discussion

❶ List the 13 safety precautions that should be taken when operating small engines.

❷ What are the four events that must take place for an internal combustion engine to run smoothly?

❸ Describe the difference between TDC and BDC.

❹ What is the main difference between a two-stroke cycle engine and a four-stroke cycle engine?

❺ List five characteristics of four-stroke cycle engines as compared to two-stroke cycle engines.

❻ What is the difference in the fuel sources of gasoline and diesel engines?

❼ What are the eight engine systems?

❽ List and describe the two types of cooling systems.

❾ What is the purpose of a lubrication system?

⑩ Why do you think proper small engine operation is important in agriculture?

CHAPTER 23

Biotechnology: The Future of Agriculture

Student Objectives

When you have finished studying this chapter, you should be able to:

- Relate why agriculture will always exist.
- Cite examples of how biotechnology has been used in the past.
- Explain the current uses of genetic engineering.
- Discuss some future uses of genetic engineering.
- Explain the concerns over genetic engineering.
- Discuss careers in agricultural science.

Key Terms

biotechnology	embryo transfer	genetic engineering
artificial insemination	clone	

417

FIGURE 23–1
Since ancient times many changes have occurred in the way humans grow food. *Courtesy of James Strawser, The University of Georgia.*

Over the years many industries and occupations have come and gone. As jobs outlive their usefulness, the industry dies out. One industry that can never die out is agriculture. People must have food, shelter, and clothing. All of these needs, especially food, depend on agriculture.

Even though agriculture will always exist, the way things are done in agriculture will change. Just look at all of the changes that have occurred since humans first began to till the earth using sharpened sticks. Although changes took place over thousands of years, most have occurred during the last century (Figure 23–1). Futurists tell us that the changes you will see during your lifetime will be greater than any that have occurred so far. That is hard to imagine!

Biotechnology

The most profound changes in the future will occur in **biotechnology.** This term refers to the application of biological and engineering techniques to living organisms. This concept is not new. In fact, biotechnology has been in use for thousands of years. Perhaps the earliest biotechnology was used to make cheese and wine. Grape juice and milk spoiled very rapidly. Through the use of certain types of bacteria, the milk could be fermented to form cheese and the juice could be fermented to form wine (Figure 23–2). The use of these bacterial agents transformed highly spoilable foods into foods that could be kept for a long period of time. In the same manner, food such as fresh cabbage could be fermented to make sauerkraut that would last longer.

Ancient biotechnology was not limited to food preservation. Legend has it that hundreds of years ago, Arabs first used **artificial insemination.** These people were noted for the beautiful horses they grew. According to the legend, Arabs from one tribe would sneak into the camp of enemy tribes and collect semen from the prized stallions of the enemy chieftains. They would then use the semen to impregnate their mares. This artificial mating produced fine horses.

Reproductive Technology

Reproductive technology has progressed dramatically since the time when the Arabs first began artificial insemination. Both artificial insemination and **embryo transfer** are now commonplace technologies. Both dams and sires are chosen through the use of computers. The ability to freeze and store semen and embryos has tremendously affected the production of superior animals (Figure 23–3).

Scientists can now even **clone** animals. Cloning is the ability to produce an organism that is genetically identical to another. Plant scientists have been able to clone plants for some time using a technique called tissue culture. Animals have been more difficult to clone. However, several companies now offer frozen embryos for sale that have been cloned from superior parents.

Another reproductive technique that is just around the corner is the development of all-male or

FIGURE 23–2
Long ago grape juice would spoil unless it was fermented into wine. *Courtesy of James Strawser, The University of Georgia.*

FIGURE 23–3
The use of artificial insemination and embryo transfer has greatly advanced the animal industry. *Courtesy of James Strawser, The University of Georgia.*

FIGURE 23–4
Dairy producers place a much greater value on female calves than they do on male calves. *Courtesy of James Strawser, The University of Georgia.*

#16

all-female embryos. This technology would be valuable to producers who only produce one or the other sex. For example, if a producer is in the business of hatching pullets to be grown into laying hens, only the females are valuable. Now, about half of the eggs that hatch are male. If the process could be developed to produce only eggs that would hatch female chicks, the efficiency of the hatchery could be doubled. Another example would be the production of heifers for dairies. Bull calves are not nearly as valuable as female calves that will one day produce milk (Figure 23–4). The technology already exists to produce embryos that are only male or only female. The challenge is to make the process economically feasible.

Genetic Engineering

The newest form of biotechnology is **genetic engineering.** Genetic engineering is the altering of the genetic makeup of organisms. Genes are the substances that determine the characteristics of an individual organism. One of the first uses of this new technology was the manufacture of human insulin.

This substance is produced by the human body and helps to regulate the metabolism of carbohydrates. If the body does not manufacture enough insulin, a condition known as diabetes results. Previously, insulin to treat diabetes was harvested from the pancreases of slaughtered cattle and pigs. Because only a tiny amount could be obtained from an animal, the process was expensive. Also, cattle insulin was not exactly the same as human insulin and some people were allergic to it.

This problem was solved through the use of genetic engineering. Scientists were able to find the gene part that regulates the production of insulin. They then spliced this segment into the gene of the E. coli bacteria. The bacteria that produce insulin multiply and pass the capability along to the next generation. Because the genetically altered bacteria are weaker than the ordinary bacteria, they have to be grown under carefully controlled conditions.

When the proper number of bacteria is reproduced, they are taken apart to retrieve the insulin produced by the bacteria. The insulin is purified, and the remains of the bacteria are destroyed. The genetically altered bacteria produce a ready, relatively inexpensive supply of insulin.

Bacteria have become the manufacturing centers for many substances that have made the lives of humans better and more productive. Vaccines that are used to impart immunity into animals and humans are produced in much the same way as insulin. Hormones that control growth and other bodily functions are also produced using genetic engineering.

Through genetic engineering, scientists have developed plants that are resistant to herbicides. Herbicides are chemicals that are used to kill weeds or unwanted plants. When the weeds are growing among a crop, a herbicide must be used that will kill the weeds and not harm the crop.

_# 17_

Genetic engineering can develop crop plants that are tolerant of herbicides. This is a tricky process because the genetically manipulated crop cannot be guaranteed to retain all of the characteristics of the nonmanipulated crop. For example, grain that is grown for food still has to produce well and also must contain the proper amounts of nutrients.

Career Development Events

The FFA Agriscience student recognition program not only rewards the students, but it is designed to educate parents, school personnel, and the public about the career opportunities available in agriscience. The program recognizes students who are pursuing challenging school-based courses focused on the scientific principles and new technologies of agriculture. The winners of this award receive financial assistance to attend college. This helps to produce a reliable supply of agriscience graduates to meet society's future needs.

To be eligible for participation in the agriscience recognition program, the student must be an FFA member who is a high school junior or senior or a first-year student in college. The student must be taking academically demanding courses including classes in agriculture and be planning a career in agriculture that requires at least a bachelor's degree. The student must submit an application and list of agriscience activities and completed projects. These activities will be judged by awarding points for skill, scientific thought, thoroughness, clarity, and creative ability. The student's academic achievement and school and community activities are also judged.

The student's agriscience project must follow the scientific research process. Each phase of the project must be identified, explained, and recorded. Tables, lists, or graphs are a good way to illustrate observations made during the experiment. The agriscience project should fall into a plant science, animal science, natural resources, earth science, energy production and management, food science, or engineering science category. Applications should include letters of recommendation from an employer, instructor, counselor, or other school official plus pictures showing the student participating in the project.

Scientists have developed the ability to determine exactly where the part of the gene is located that controls certain characteristics. They can do this for some characteristics on some organisms. Just think of the possibilities when they are able to map out and identify exactly where every characteristic is located for every organism. For example, bacteria

The Agriscience Student Recognition Program must be submitted through the agriculture instructor. The state winner will be presented a scholarship for use at the college of choice and is eligible to compete for regional and national awards. Regional and national winners can win scholarship money that totals over $8,000.

The Agriscience Award recognizes students who have excelled in the science of agriculture. *Courtesy of Blane Marable, Morgan Co. High School, Madison, GA.*

live in a ruminant's stomach and allow the animal to digest large amounts of fiber. One day scientists may be able to genetically alter the bacteria to make them able to digest wood fiber. This would allow cattle and other animals to be fed on woody fiber that is now going to waste. Perhaps monogastric (simple-stomached) animals such as pigs could be developed that could digest roughages like ruminants. This would mean a much cheaper way to produce pork.

Another possibility may be to develop plants like corn that produce their own nitrogen. Corn requires a lot of nitrogen in order to produce large yields. This fertilizer is expensive to supply. Some plants, called legumes, produce nitrogen through the use of bacteria that live in nodules on their roots. If scientists could genetically alter the corn plant (along with other grasses) to allow it to host the nitrogen-producing bacteria, grain could be produced much more efficiently (Figure 23–5).

Crops may also be altered to produce more oil or oil of a better type. Remember from Chapter 10 that some crops are produced for the oil from their seed. Scientists think that in the future these as well

FIGURE 23–5
If scientists could alter corn to make it a legume, much less would be spent on fertilizer. *Courtesy of Dekalb Genetics Corporation.*

as other crops may be altered to produce a much higher percentage of oil. This may go a long way in helping to solve the problems caused by running out of petroleum.

Genetic engineering has the potential to be revolutionary in the production of food and fiber. Scientists estimate that early in the twenty-first century, there will be over six billion people in the world. All of these people will have to be fed and clothed. Through the use of genetically altered plants and animals, the efficiency of producing food can be increased and areas that are currently of little agricultural value can be made productive.

Even though there is much hope and optimism over genetic engineering, there is also widespread concern. Some people are worried about the introduction of genetically altered organisms into the environment. They cite examples of plants and animals that were introduced with good intentions and today have escaped and caused problems. They feel that by tampering with nature we may unleash organisms that will be hard to control. Many science fiction movies depict scary creatures that have been genetically altered. People note that the most commonly used organism in gene alteration is the E. coli bacteria that normally live in the human intestinal tract. Because the bacteria multiply rapidly and the human body is a host for these bacteria, many people feel that we do not know enough about genetic alteration to allow these bacteria into the environment.

The answer to this dilemma is, at least in part, the regulation of genetic engineering. Scientists conduct research on genetic engineering in closely controlled laboratory situations. The buildings and equipment are designed to prevent the escape of newly engineered microorganisms into the environment (Figure 23–6). Research may take many years before the determination is made that the organisms will be tested outside the laboratory.

FIGURE 23–6
Facilities used in genetic engineering are designed to prevent the escape of organisms. *Courtesy of USDA ARS.*

The United States Department of Agriculture (USDA) has developed strict guidelines for testing genetically altered organisms that can come in contact with the environment. In addition, the USDA has developed a set of guidelines that all USDA research agencies will use prior to field testing. A heightened public awareness helps to ensure that those doing research remain diligent and do not become careless. It also helps everyone remain conscious of the fact that problems could result from carelessly conducted genetic research.

Careers in Agriscience

You may not have thought about what sort of work you want to do, but sooner or later you must decide. Consider a career in the nation's largest industry—agriculture (Figure 23–7). From stories in the media, you may think that agriculture is dying out. Nothing could be further from the truth. Agriculture is not dying—it is changing. The changes are exciting and challenging. Remember that all the people in the world will have to be fed and clothed. This will be done through agricultural science.

FIGURE 23–7
Perhaps you should consider a career in the nation's largest industry. *Courtesy of Patrick Smith, The University of Georgia.*

FIGURE 23–8
Agriculturists work in a variety of settings. *Courtesy of USDA ARS.*

Consider all the things you like to do. A substantial part of your life will be spent working, and if you enjoy your work, your life will be more pleasant. If you like to work with plants or animals and if you like science, a career in agriculture might be for you. Some people have the mistaken idea that a career in agriculture is difficult, strenuous, dirty work that has no challenge. Today's agriculture can offer exciting challenges. You may work in a variety of places ranging from a food processing plant to a laboratory (Figure 23–8).

Summary

As with any other career, you will need the proper preparation for a career in agriculture. The best place to begin is with the high school agricultural education program. In this program you will decide which areas you wish to specialize in. Throughout this book are examples of activities you can get involved in through the FFA. The opportunities are limitless. You can even earn scholarships to help pay for a college education. Agriculture is taught as a two-year program in community and technical colleges, or as part of a four-year university program. These institutions will be glad to send information on opportunities in all their degrees and programs.

23 CHAPTER REVIEW

Student Learning Activities

1. Think of a problem facing agriculture today or in the future. Write a paper giving your ideas as to how the problem could be solved using biotechnology. Be sure to include possible hazards.

2. Interview five people and get their perceptions of genetic engineering. Do you feel their perceptions are correct? Report to the class on your conclusions. Be sure to keep the identity of the people you interviewed confidential.

3. Research the career opportunities in biotechnology. Determine the job responsibilities of a particular career and the qualifications needed. Report to the class on your findings.

4. Search the Internet and find five positive aspects of genetic engineering and five negative aspects. Which aspects are the most credible?

True/False

1. ___ Most changes in agriculture occurred during the twentieth century.

2. ___ Dams and sires are chosen through the use of computers.

3. ___ Scientists have not yet discovered how to clone animals.

4. ___ The majority of insulin is harvested from the pancreases of cattle.

5. ___ Corn requires a lot of nitrogen in order to produce large yields.

6. ___ Scientists estimate that early in the 21st century there will be over six billion people in the world.

⑦ ___ Legumes produce nitrogen through the use of bacteria that live in nodules on leaves.

⑧ ___ The nation's largest industry is agriculture.

⑨ ___ Agriculture is not dying; it is changing.

⑩ ___ A career in agriculture may involve working in a variety of places such as a food processing plant or a laboratory.

Multiple Choice

① One industry that can never die out is
    a. agriculture.
    b. computer production.
    c. engineering.

② Perhaps the earliest biotechnology was used to make wine and
    a. bread.
    b. apples.
    c. cheese.

③ One of the newest forms of biotechnology is
    a. insecticide production.
    b. manufacturing herbicides.
    c. genetic engineering.

④ One of the first uses of genetic engineering was in manufacturing
    a. cotton.
    b. human insulin.
    c. pesticides.

⑤ Genetic scientists have been able to develop plants that are resistant to
    a. herbicides.
    b. biotechnology.
    c. predators.

⑥ Bacteria live in a ruminant's stomach and allow the animal to digest large amounts of
    a. fiber.
    b. biotechnology.
    c. grain.

7 The best place to begin preparing for an agricultural career is with an agricultural education program in
   a. preschool.
   b. elementary school.
   c. high school.

8 The most profound changes in the future will occur in
   a. building construction.
   b. genetic engineering.
   c. animal care.

9 Food, shelter, and clothing are all directly dependent on
   a. agriculture.
   b. crops.
   c. water.

10 Substances that control growth and other bodily functions are
   a. vaccines.
   b. hormones.
   c. herbicides.

Discussion

1 Why is cattle insulin no longer used to help diabetics?

2 Why do scientists want to genetically alter the bacteria found in a ruminant's stomach?

3 What impact does genetic engineering have on plants?

4 Why is there concern about genetic engineering?

5 How is genetic engineering being regulated?

6 How can you prepare for a career in agriculture?

7 Why will agriculture always exist?

8 What are some current uses of genetic engineering?

9 What are some future uses of genetic engineering?

10 How was biotechnology used in the past?

CHAPTER 24

High School Agricultural Education Programs

Student Objectives

When you have finished studying this chapter, you should be able to:

- List the advantages of studying Agricultural Education in High School.
- Name the three parts of the Agricultural Program.
- Describe the different types of Supervised Agricultural Experience programs.
- Discuss the mission of FFA.
- Discuss how Agricultural Education is different from the old Vocational Agriculture Programs.

Key Terms

Agricultural Education

Vocational Agriculture

Supervised Agricultural
  Experience

National FFA
  Organization

Career Development Event

Proficiency Award

Modern Programs in Agriculture

If you have enjoyed studying the chapters in this book, you will probably want to consider studying more about Agriscience. When you enter high school, an entire program centered on the world of agriculture will be available to you. In every state in the country, there are local programs of **Agricultural Education** designed to help you understand agriculture and to help you develop many different types of skill.

For over 80 years, these programs have helped high school students learn about themselves and the world all around them. The programs began in 1917 and were designed for rural students who lived on a farm. Back then, the program was called **Vocational Agriculture.** The curriculum and activities were designed to help solve problems faced on the farm. The program also helped students to develop life skills such as social abilities. Over the years, the program has changed a lot. It is no longer a program aimed at students who live on farms (Figure 24–1).

FIGURE 24–1
Agricultural Education is not just for students who live in rural areas.
Courtesy of USDA

In fact, according to the National FFA Organization, more students enrolled in the program live in urban or suburban areas than live in rural, farm areas. Even the name has changed. Several years ago, the name was changed from Vocational Agriculture to Agricultural Education. The new name better described the more modern activities. Agricultural Education is made up of three parts—classroom and laboratory activities, **Supervised Agricultural Experience** (SAE), and FFA. In the classroom and laboratory, you will learn about a wide variety of topics. The classroom instruction will introduce you to an area of study. Here you may learn about the theory behind a topic. In the lab, you will have hands on activities to learn the practical applications. Popular areas of study include forestry, horticulture, Agriscience, wildlife and natural resources, as well as several other areas. Classroom and lab study in Agricultural Education may be a little different from what you are used to. Here you will get an opportunity to actively participate in learning. This means that you will learn by doing. The laboratory may be a little different than what you think of as a lab (Figure 24–2).

The agriculture lab may be a greenhouse, a well-equipped mechanics laboratory, or a forestry plot. It may be a pond where you learn about raising fish or it may be a nursery where you learn to propagate plants. You will grow plants in the greenhouse, take and analyze water samples, estimate the amount of lumber in a growing tree, grow bacterial cultures, taste dairy products, or wire a light switch. Each day you will have different and interesting activities to do. The science and math concepts learned in other courses will come alive in Agriculture class!

Supervised Agricultural Experience

Learning about the exciting world of agriculture does not end when the school day is over. Through

FIGURE 24–2
The Agricultural Education classroom and lab may be different than what you expect. *Courtesy of Dr. Frank Flanders, Georgia Department of Education.*

FIGURE 24–3
Your teacher will work with you after school in a Supervised Agricultural Experience Program (SAEP). *Courtesy of Dr. Frank Flanders, Georgia Department of Education.*

the Supervised Agricultural Experience Program (known as SAE or SAEP) component of Agricultural Education, you can be involved in interesting activities after school. One of the best parts is that you can pick the activity that interests you. Your teacher will assist you in determining the correct program for you. Once you have decided, your teacher will help you get started and visit with you periodically (Figure 24–3).

During the visits, he/she will help you to solve any problems and provide any needed assistance.

Types of SAEs

There are four basic types of SAE:

Exploratory—This type allows you to explore different areas of interest. For example, you might want to investigate the duties of a veterinarian. You might do research through the Internet, read articles about the profession, or talk to veterinarians about what they do on the job (Figure 24–4).

The next step would be to visit with the veterinarian on the job and maybe even assist in some of

FIGURE 24–4
In an Exploratory SAEP, you might work with a veterinarian to learn about the duties he/she performs. *Courtesy of Getty Images.*

the duties. This will give you an idea about whether or not this would be a career for you. Exploratory Supervised Experience Programs are very broad based. You agriculture teacher will be open to almost any idea you may have for further exploring the world of agriculture.

Research/Experimentation—For your SAE, you may want to learn to do scientific research. Your teacher knows how to set up experiments and can teach you how (Figure 24–5). You will learn to use the scientific method.

You will also learn how to communicate the findings from your research. This type of SAE can be a very small experiment or it can grow into research that you conduct all the time you are in high school.

Ownership/Entrepreneurship—You may wish to start an enterprise of your own. Many successful businesses began as SAEs in high school agriculture programs. You might begin by operating a lawn-mowing business. This may develop into a full-blown lawn maintenance business. It might even evolve into a landscaping enterprise (Figure 24–6).

Many students own livestock. They might begin with a heifer, gilt, or filly and develop a livestock business. With this SAE, you have the potential for earning money when you are in high school.

Placement—There are many places where you can obtain experience with people in agriculture. You may even get paid for working in a part time job where you can gain the experience. However, placement SAEs are much more than part-time employment. Although you may earn money, the main objective is to gain experience. You might work in a nursery, on a horse farm, in a tractor dealership, or many other places (Figure 24–7). Your teacher can help you set up the placement SAE and help you design the experience.

FIGURE 24–5
In a research SAEP, your teacher will help you set up and conduct scientific research. *Courtesy of Agricultural Research service.*

FIGURE 24–6
With an ownership/entrepreneurship SAEP you might own a lawn care or landscape business. *Courtesy of Getty Images*

FIGURE 24–7
A placement SAEP might involve working on a horse farm to learn about caring for and managing horses. *Courtesy of Getty Images*

The FFA Organization

Throughout this text there is an insert in every chapter that discusses some aspect or program of the FFA. The FFA, which began in 1928, was known as the Future Farmers of America. In 1988, the delegates to the national convention in Kansas City voted to change the name to the **National FFA Organization.** This new name better represented the broader programs of the agricultural education program. The delegates felt that the organization was for students interested in all aspects of agriculture. As a result of the changes in the program and the name, the membership has continued to grow.

Over the years, the FFA has been legendary in providing opportunities for young people. The organization is much more than a means to learn about agriculture. In the FFA you can learn a lot about personal development (Figure 24–8). This means that you will be able to travel, meet many people, learn to communicate, and develop leadership skills.

Many famous people credit FFA with teaching them basic skills. For example, former President

FIGURE 24–8
In FFA you will be able to travel, meet many people, learn to communicate, and develop leadership skills. *Courtesy National FFA Organization.*

Jimmy Carter was a member in Georgia. Speaking about the FFA, he said, "I began to learn how to make a speech. And I began to learn how to work with other people. I also learned the value of agriculture farm families, stability, commitment, idealism, hope, truth, hard work and patriotism from the FFA" (National FFA Organization). Jim Davis, who writes the Garfield cartoon, was an FFA member in Illinois. Senator Sam Brownback from Kansas was a national officer in the FFA. Auburn running back and Heisman Trophy winner Bo Jackson was an FFA member in Alabama.

Today the FFA is a national organization with almost one-half a million members. There are members in all 50 states as well as in Puerto Rico and the U.S. Virgin Islands. The National FFA Organization lists the following characteristics of membership:

In 2004, there were 461,043 FFA members, aged 12–21, in 7,310 chapters in all 50 states, Puerto Rico, and the Virgin Islands

91 percent of FFA members are in grades 9–12; 4 percent are in grades 6–8; 5 percent are high school graduates

27 percent of FFA members live in rural farm areas; the remainder live in rural nonfarm (39 percent), urban, and suburban areas (34 percent)

FFA chapters are in 10 of the 15 largest cities, including New York, Chicago, and Philadelphia

The top five membership states are California, Texas, Georgia, Oklahoma, and Ohio

The 75th National FFA Convention was host to more than 51,000 members, advisors, and supporters

FFA Programs

The National FFA Organization is dedicated to making a positive difference in the lives of young people by developing their potential for premier leadership, personal growth and career success through agricultural education. This statement best describes what the FFA is all about. You will make friends and learn about working with others. All these things can be accomplished by participating in the many programs of FFA.

The FFA has several different types of programs. These include chapter activities, team activities, and individual activities (Figure 24–9). Most of the activities involve competitions on the local, state, and national levels.

Career Development Events

The National FFA Organization offers a wide variety of leadership opportunities for students to get involved in and provides many ways to explore agriculture. The FFA Global Programs offer students and teachers the opportunity to explore international agriculture through programs that teach about values, traditions, lifestyles, and the agriculture of other countries. FFA Global offers members a variety of educational experiences ranging from travel seminars to host family stays and international internships.

Traveling and learning in the FFA increases awareness of agricultural education in other countries when providing opportunities to develop programs and leadership. As students become familiar with cultural issues, they gain a better understanding and appreciation of others within their country. There are many ways that students and local chapters can get involved in Global FFA such as: chapter-to-chapter exchanges, group travel seminars, homestay program through National FFA, conferences or expositions, hosting a foreign student through an FFA International Inbound Program, or develop an international component to SAE.

International experiences can extend after high school as well. As students prepare to continue their studies and future careers, they can become immersed in global issues through a variety of opportunities. International

FIGURE 24–9
Chapter activities involve everyone in the chapter and may include competition with other chapters. *Courtesy National FFA Organization.*

The FAA Global Programs offers students the opportunity to experience lifestyles, traditions, and agriculture in different countries. *Courtesy of Getty Images.*

knowledge and experiences can be gained by internships in international areas around the world (USDA-FAS Internship Program), International Leadership Seminars for State Officers (ILSSO), World Experience in Agriculture (WEA), seminars or homestay programs, and presentations on past international experiences to educate others.

More information on international opportunities for students and teachers can be found online at http//www. ffa.org/programs/ global/index.html.

Chapter activities involve competitions such as the scrapbook competition. The National Chapter Award allows your chapter to compete with other chapters in your state and in the nation. The competition is based on the activities in which your chapter is involved.

Most team activities are part of what is known as **Career Development Events** (CDEs). These events allow you to use the knowledge and skills you have learned in the classroom and in your SAE. There is team competition in 23 different areas ranging from livestock evaluation, dairy products, forestry, and nursery/landscape. Not only can you learn a lot about the skill area, but also you can learn a lot about teamwork and getting along with others (Figure 24–10).

In addition, you will get to travel around the state and to the national convention in order to compete.

Individual activities are those you participate in as an individual person. For example, there are 45 different proficiency award areas. A **Proficiency Award** is based on your SAE program. Once you decide your area of interest in a Supervised

FIGURE 24–10
In FFA there are 23 different areas of team competition. *Courtesy National FFA Organization.*

Agricultural Experience program, you can take the program as far as you can. If you excel, you can receive an award for your accomplishments. Proficiency awards carry some nice monetary prizes for the winners.

Summary

No matter what career path you may choose, the Agricultural Education program can help you gain skills that will help you achieve your goals. The three parts of the program (classroom/lab, SAE, and FFA) will provide you with many opportunities to learn and apply knowledge and skills. Experiences you have in FFA will lead to skills that will be of benefit to you no matter what you do for a career. Don't forget that one of the main advantages of the Agricultural Education program is that it makes learning fun.

24 CHAPTER REVIEW

Student Learning Activities

1. Make a list of the topics in this text that interest you the most. Share your list with the class. Explain why these topics interest you.

2. From the list you made in the first activity, design an SAE. What type of SAE is best for you? What do you expect to learn from the experience?

3. Visit the FFA Web site at http://www.ffa.org. What activities appeal to you?

4. Visit a high school Agricultural Education program. Ask the teacher to explain the activities of the program. Find out which of these activities appeal to you.

True/False

1. ___ Only a few states still have high school agricultural programs.

2. ___ More students enrolled in Agricultural Education live in urban/ suburban areas than in rural areas.

3. ___ The name of the program is called Vocational Agriculture.

4. ___ The name of the Future Farmers of America has been changed to the National FFA Organization.

5. ___ Three parts of the program are classroom/lab, SAE, and FFA.

6. ___ Supervised Agricultural Experience (SAE) programs include only those activities related directly to farming.

7. ___ Agriculture Teachers work with their students after the school day has ended.

8. ___ One of our nation's past presidents is a former FFA member.

9. ___ The top membership states are California and Texas.

10. ___ Agricultural Education has programs of interest to almost everyone.

Multiple Choice

1. The agriculture lab may be a:
   a. well equipped mechanics laboratory
   b. a green house
   c. both a and b

2. The Future Farmers of America was organized in:
   a. 1933
   b. 1928
   c. 1917

3. Which former US President was an FFA member?
   a. Herbert Hoover
   b. Bill Clinton
   c. Jimmy Carter

4 FFA began in:
- a. St. Louis
- b. Louisville
- c. Kansas City

5 There are FFA members in:
- a. all 50 states as well as Puerto Rico and the Virgin Islands
- b. only states with large rural populations
- c. 43 of the 50 states

6 Exploratory, Research, Ownership, and Placement are types of:
- a. FFA programs
- b. Classroom activities
- c. SAE

7 SAE stands for:
- a. standard agricultural event
- b. supervised agricultural experience
- c. special agricultural event

8 An award based on the activities in which your chapter is involved is called:
- a. Career development Award
- b. National Chapter Award
- c. Team activity Award

9 Agricultural Education was once called:
- a. Vocational Agriculture
- b. Agriculture Programs
- c. Studies in Agriculture

10 _____ are based on your SAE program.
- a. National Chapter Awards
- b. Proficiency Awards
- c. Career development Events

Discussion

1 What are the three parts of an Agricultural Education Program?

2 Name three types of laboratories used in Agricultural Education.

3 Name the four types of SAEs.

4 Why did the Future Farmers change its name to the National FFA Organization?

5 Name three famous people who are former FFA members.

6 How many FFA members were there in 2004?

7 What is a Career Development Event?

CHAPTER 25

Careers in Agricultural Science*

Student Objectives

When you have finished studying this chapter, you should be able to:

- Identify the main job classifications in agriculture.
- Analyze the career opportunities in agriculture.
- Identify some of the characteristics associated with various jobs in agriculture.

Key Terms

agribusiness	horticulturists	food scientists
agronomists	geneticists	microbiologists
entomologists		

---

*This chapter was contributed by Jean Kilnoski.

FIGURE 25–1
More than 20 million workers are employed in processing, storing, and distributing food and fiber. *Courtesy of Henry Waeti.*

During the first half of the twentieth century, agriculture was mostly limited to farming, and preparing for a career in agriculture was fairly easy. People did not think that a formal education was necessary for a farmer. Today, successful people who work in agriculture need a good education in business and in science. Over the years, the number of farmers in the United States has declined dramatically, yet agriculture remains a major source of jobs for Americans. Agriculture professionals are not just farmers and ranchers. Opportunities in this field include areas as diverse as science, engineering, finance, marketing, and exporting. Careers in these areas have expanded as the production and processing of food and fiber have become more specialized and technological. Successful producers rely heavily on the support of these workers. Individuals working in agriculture today are as likely to work in the city as on the farm.

The business of agriculture, or **agribusiness,** is an important part of the U.S. economy. More than 20 million workers are employed in processing, storing, and distributing the food and fiber produced by farmers (Figure 25–1). Agribusiness also involves manufacturing and selling farm supplies and equipment. The continued success of agriculture depends on the support of this entire network of people who enable the producer to function more efficiently. It is their initiative and hard work that make American agriculture competitive in the world market.

Plant Science

Plant science deals with the growing of fruits, vegetables, and ornamental plants. Career opportunities in plant science vary widely, but all of them are involved in producing plants. The main categories in plant science are horticulture, agronomy, landscaping, and turf management.

FIGURE 25–2
Crop scientists work to increase the yield of crops by improving new plant strains. *Courtesy of USDA.*

The production of field crops requires the expert knowledge of many scientists. **Agronomists** provide producers with information about the combination of soils, plants, and climate that will promote the best possible yield. **Entomologists** provide information about pest control. Today insect populations are not always controlled with pesticides. Parasitic or predator insects can be released in a field to control a pest population. Crop scientists work to increase the yield of crops by improving farming methods and developing new plant strains. Some crop scientists study the breeding of crops and use genetic engineering to develop crops resistant to pests and drought (Figure 25–2). All of these agricultural scientists must have advanced college study in their chosen fields.

Horticulture deals with the cultivation of fruits, vegetables, flowers, trees, and shrubs. Some **horticulturists** find new ways to grow plants or breed better varieties. They experiment with different varieties of fruits, vegetables, and nuts to make them more nutritious. Others raise and tend plants used for ornamental purposes, such as flowers, shrubs, trees, and grasses. An increase in the interest for indoor and outdoor plantings has raised the demand for these services. People want attractive grounds around their homes and offices, as well as amusement parks, gas stations, fast food stores, and shopping malls. Skilled horticulturists are needed to grow and tend plants in all of these areas.

Many jobs in plant science require two years of study or less (Figure 25–3). Nursery operators raise and sell flowers, shrubs, hedges, and trees. They are involved in tasks including plant propagation through seeds and cuttings, preparation of soil in outdoor growing areas, greenhouse management, and the storage and packing of plants.

Landscape architects design outdoor areas for people to use, such as parks or gardens, housing projects, or the grounds surrounding hotels. They plan their arrangements to make the best use of the land while protecting the natural environment. They prepare detailed maps and plans showing the placement of trees, shrubs, and walkways. The landscape architect also supervises any grading or construction that is necessary to complete the work.

FIGURE 25–3
Many careers in turf production/ maintenance and landscaping require less than a four-year college degree. *Courtesy of Progress Farmer Magazine.*

Turf grass technicians may work in commercial lawn maintenance, private lawn care, or the maintenance of parks, highways, and playing fields. They may be employed as greenskeepers for golf courses, work for the government maintaining public lands and parks, or work in their own businesses. They provide numerous services that include mowing and fertilizing, as well as insect, weed, and disease control.

Animal Science

Careers in animal science are involved with many phases of animal production. These professions include producers, feeders, marketers, manufacturers, distributors, and retailers. They may also include producing pharmaceuticals or equipment that aids in the management of animals. Many of these careers are highly science oriented. The intensive health practices that are necessary to maintain a successful operation in cattle ranching or dairy farming provide many job opportunities in animal welfare. Scientists study the nutrition, growth, and development of domestic animals (Figure 25–4).

FIGURE 25–4
Animal scientists study the nutrition, growth, and development of animals. *Courtesy of The University of Minnesota Agricultural Experiment Station.*

Geneticists look for ways to develop high growth-rate animals that result in a quality product that meets consumer standards. These positions require a four-year program of college study that emphasizes the sciences.

Cattle reproduction has opened the door to another career in the animal sciences. Artificial insemination is the most common cattle-breeding practice today. Artificial insemination technicians typically have two years of college education. Feedlots also present career opportunities in animal

Career Development Events

The National FFA Convention was founded in 1928, and until 1999 FFA members from all over the country met in Kansas City. In 1999, the convention was moved to Louisville, KY. The convention has grown to become the world's largest student educational convention. Students from every state in the country, plus students from several foreign countries, attend. All types of career development events are conducted during the convention ranging from public speaking to agriscience. Tours to educational sites are conducted almost every day. Students have a wide array of leadership development workshops to choose from where they can interact with students from all over the country.

At the convention, students receive awards for many activities that began on the local level and progressed through to the national level. It is quite a good feeling to know that they have earned a national award through hard work and dedication. The FFA awards over a million dollars in scholarships each year.

One of the many highlights of the convention is the gigantic Careers Show. Companies from all over the world set up booths to tell about their company. Students can learn about thousands of job opportunities involved with agriculture. Many colleges from all across the United States operate booths where students can get information about various programs and scholarships. They can go home with stacks of printed materials about job opportunities.

Most of the students attending the convention have earned their way. As

science. At the feedlot the cattle are fattened up so that their market value improves. The manager of a feedlot is responsible for minimizing the costs of feeding the cattle in order to obtain the highest profit. Feedlot operations use computers to regulate feed supplements and medical products. Many feedlot managers are producers who grow their own hay and grain.

Cooperative extension agents provide services to both producers and apartment dwellers. Agents are able to advise people on topics as diverse as crop

you begin to study agriculture in high school, set your goal to attend the National FFA Convention. Your teacher can help you design a plan to achieve your goal. Remember, the opportunities in FFA are limited only by your ambition and willingness to set goals and work toward them.

The National Career Show at the National FFA Convention provides information for students on hundreds of different careers in agriculture. *Courtesy of Bill Stagg, National FFA Organization.*

and livestock production, conservation, home economics, child care, farm management, and nutrition. Cooperative extension agents can answer questions about when to transplant seedling tomatoes or how to remove stains from clothing. Agents often specialize in one of these areas. Extension agents are employed jointly by county and state governments. They must have a four-year degree in one of the fields that relates to extension work.

If you enjoy being around animals you might want to consider pursuing a career in one of the many areas dealing with companion animals. The most important attributes you need for a career in companion animal care are respect and love for animals.

Pet care workers provide a wide variety of services to the owners of small animals. They work in animal hospitals, boarding kennels, and animal shelters. Responsibilities will vary, but may include feeding animals and cleaning and disinfecting cages and equipment. Pet care workers can also work in pet grooming parlors, pet training schools, and pet shops. Many of these positions are entry level and require no special education. Employers often hire high school students and graduates to fill these positions. Some experience with animals is helpful.

Animal trainers teach animals to obey commands, compete in shows or races, or perform tricks. Many companion animals can be trained, including dogs, horses, parakeets, and cockatoos (Figure 25–5). Animal trainers can teach dogs to perform a variety of tasks for humans, including the protection of property. Employees in this field may also be involved in training assistance dogs for the blind, the deaf, or the physically challenged. Most animal trainers begin their careers as pet care workers where they gain experience caring for animals.

FIGURE 25–5
People who train animals are also in an agriculturally related field. *Courtesy of Kentucky Horse Park.*

Dog groomers wash, comb, cut, and trim the fur of dogs and cats. To become a dog groomer it is best to take a course from one of the 50 schools accredited by the National Dog Groomers Association. To enroll in the school, you must be at least 17 years old. Some animals may be nervous, fidgety, or aggressive, so patience and persistence are necessary characteristics of a potential dog groomer.

Horse industry workers breed, train, and care for horses. People who work with racehorses usually begin their careers as grooms. Grooms are responsible for brushing and combing the horses. They are often in charge of feeding the animals and cleaning the stables where they are kept. Grooms may also be involved in training young horses.

Horse trainers are responsible for determining the best diet and exercise program for the horses. The trainers often supervise grooms and other employees who exercise and care for the horses. Nearly all horse industry workers learn at least some of their skills on the job.

Animal health technologists work side by side with veterinarians. They assist veterinarians with laboratory work, radiology, anesthesiology, and surgery (Figure 25–6). They may also be responsible

FIGURE 25–6
Animal health technologists may
assist with laboratory work.
Courtesy of USDA.

for animal nursing, sample collection, and general office management. An animal health technologist is not limited to work in a veterinary clinic or laboratory; he or she may also work in a zoo or a circus. These technicians are graduates of a two-year program from a community or technical college.

Veterinarians are animal doctors. They perform a variety of duties in a number of different environments. Veterinarians diagnose animal illnesses, treat diseased and injured animals, perform surgery, and prescribe and administer drugs and vaccines. They also give animal owners advice on the care and breeding of animals. They can work in a clinic, conduct research for the government, or work for a large organization. They can treat both large and small animals, or they can specialize. Almost one-third of all veterinarians specialize in the care of small animals.

Veterinarians must complete four years of college followed by an additional four years of study at a veterinary school. To become licensed to practice, veterinarians must also pass their state's oral and written licensing examination.

Food Science

The food industry is one of the largest industries in the nation. It offers a wide range of opportunities because of the variety of operations and services it performs. Only a small portion of the foods we consume are fresh. Most have been processed by the time they reach the local supermarket. Before the food product can be sold to the consumer it must be processed, inspected, packaged, labeled, stored, and transported. All of these jobs are part of the extensive processing of food products, from the time the crop is harvested until it reaches the consumer.

One of the main purposes of processing foods is to keep them from spoiling. Processed foods are canned, waxed, dried, frozen, bottled, and pickled to preserve them. Another purpose for processing is to enhance the nutritional value of the food. Highly technical methods of achieving these goals have been developed that require the skills and knowledge of **food scientists**, **microbiologists,** and food chemists. All of these positions require advanced college study.

Food scientists improve methods of freezing, canning, storing, packaging, and distributing food (Figure 25–7). They study the chemical changes that take place in food when it is stored or processed. Food scientists try to find ways to process food so that fewer nutrients are lost. They also study the effect of food additives. Many food scientists are attracted to research and development. The most thrilling task of the food scientist is the development of new products. This is a highly competitive area. Developing a product usually takes more than a year and a substantial financial investment by the processing company. Only one product out of ten will survive the first year in the supermarket. Ready-to-cook convenience foods continue to account for a major part of the new product development efforts. Thanks to convenience foods, people are able to

FIGURE 25–7
Food scientists improve methods of storing, packaging, and distributing food. *Courtesy of USDA.*

spend significantly less time in the kitchen preparing meals than they did just a decade ago. Frozen, precooked meals that come in microwaveable packages are a good example of the innovations that have occurred in the food industry in recent years.

Careers in meat processing often deal with the protection of processed meats. Microbiologists lead the way in food safety research. They help the industry find ways to prevent bacterial growth. Food chemists work in the meat industry to ensure a healthful, tasty product. They may study the chemistry of fats in chicken skins to help reduce the cholesterol in human diets or the muscle chemistry of meats that causes some cuts to be tender and others to be tough. Meat inspection is another career that focuses on a safe product for the consumer, whether it is poultry, red meat, or fish (Figure 25–8). Inspectors who work for the Department of Agriculture inspect and grade almost all commercially marketed meats.

Food technicians graduate from a two-year program of study. They work in food processing plants

FIGURE 25–8
Meat inspection is a career that focuses on a safe product for consumers. *Courtesy of USDA.*

and in laboratories. They usually help with quality control, testing, or research under the supervision of a food scientist. Technicians run tests to check processed foods for bacteria, impurities, or poisons. They use instruments that check the quality of the food, including the taste, smell, and color. Some food technicians work as production supervisors and managers in food processing plants.

A number of careers in the food industry deal with agricultural marketing. Agricultural marketing involves the storing, transporting, or marketing of food products. Supermarkets are dependent on the efficient functioning of the agricultural marketing sector. Transportation specialists use all of the forms of transportation available, including trucks, trains, and barges, to get the lowest price possible while maintaining the quality of the product. These specialists must have a good working knowledge of the state and federal laws and regulations that govern the transportation of these products.

Grain merchants are involved in all aspects of buying and selling grain. They act as a connection between the producer and the ultimate consumer. They are concerned with the quality, the shipping, and the processing of the grain. Grain merchants buy most of their grain from area producers at local grain elevators. Merchants must be knowledgeable about the current market prices for grain. Many use computers to get up-to-date prices. Although there are no specific education requirements for grain merchants, some college training is helpful.

The wholesaler is a link in the marketing chain between the producers and the retail store in which the goods will be sold. The wholesaler provides a valuable service by bringing these two segments of the marketplace together. Wholesalers also buy goods in the United States for sale abroad, as well as buying products from foreign markets for sale here.

FIGURE 25–9
Supermarket managers can be responsible for more than 15,000 food and consumer items. *Courtesy of The University of Minnesota Agricultural Experiment Station.*

The retail store manager who buys the food product from the wholesaler is responsible for every phase of the store's operation. Supermarket managers can be responsible for more than 15,000 food and consumer items (Figure 25–9). They also supervise other employees, manage the inventory, account for expenses, and maintain a good relationship with the consumer. The retail manager must constantly evaluate how well each product is doing in order to make effective decisions about when to eliminate one product from the shelf to make room for a new one. This type of evaluation is often made possible by electronic devices that read product bar codes. There are no specified educational requirements for wholesalers and retailers, but a college education is highly recommended.

Natural Resources

In recent years, people in agriculture have been challenged to find ways to grow enough crops to satisfy the needs of consumers without damaging the environment. This problem will become more

FIGURE 25–10
A career with the forestry service may deal with protecting the environment. *Courtesy of USDA.*

difficult to solve as the demand for food increases. Many people in agriculture are devoted to the protection and conservation of the environment (Figure 25–10). Highly educated and trained people will be needed to protect the nation's forests, wilderness areas, and recreations areas. Most of the careers available in this area require a four-year college degree.

Soil scientists study the characteristics of soil. They gather information and give advice on the management and conservation of soil and water as well as erosion. Many soil scientists work for the U.S. government. These scientists are called soil conservationists.

Range managers oversee the use of rangelands that have been set aside for livestock grazing and other uses. They supervise the use and conservation of rangelands owned by the government. Range managers balance the grazing needs of ranchers and conservation measures needed to protect the environment. The majority of range managers work for the government. Others may work as appraisers for banks or for companies that have large holdings of rangeland.

FIGURE 25–11
Agricultural research will become even more important in the years to come. *Courtesy of USDA.*

Another career that deals with protecting the environment is with the U.S. Forest Service. Forest Service researchers conduct studies on wildlife and fish. They study the habitat requirements of individual species to determine how certain activities in the forests and on the rangelands will affect them. The studies allow them to make informed decisions about cutting timber or allowing grazing of the rangelands.

Summary

The future of careers in agriculture looks bright. If current predictions hold true, workers will need to become even more efficient because the demand for food and other agricultural goods will continue to increase. If the world's growing population is to be fed, world food production must double in the next 40 years. In addition, changing consumer tastes will continue to prompt new product development. Agricultural research will become more important in the years to come as new technologies are needed to meet the needs of the American consumer while maintaining a healthy environment (Figure 25–11).

 CHAPTER REVIEW

Student Learning Activities

❶ Choose a career that interests you. List as many of the tasks performed in this career as you can. Find out what qualifications are required to be hired in this position.

❷ Interview a person who works in the career you have chosen. List all the things he or she enjoys about working in that job.

③ Look in the classified ads in your local newspaper. How many jobs are available that deal with agriculture? What are the qualifications for the jobs?

True/False

① ___ Agriculture is the same thing as farming.

② ___ Horticulturists may raise plants for ornamental purposes.

③ ___ Cooperative extension agents only serve producers.

④ ___ Most of the foods that we consume today are fresh.

⑤ ___ Careers in meat processing often deal with the protection of processed meats.

⑥ ___ One of the main purposes of processing foods is to alter the flavor.

⑦ ___ There will be fewer jobs in agriculture in the future.

⑧ ___ An individual looking for a job in agriculture today will probably work in the country.

⑨ ___ More than 20 million people are employed in agribusiness.

⑩ ___ Almost one-third of veterinarians specialize in the care of small animals.

Multiple Choice

① The business of agriculture is called
    a. horticulture.
    b. agronomy.
    c. agribusiness.

② Entomologists provide producers with information about
    a. pest control.
    b. the best combination of soils, plants, and climate.
    c. breeding ornamental plants.

③ Individuals who deal with the chemistry of fats in chicken skins are
    a. food chemists.
    b. microbiologists.
    c. entomologists.

④ Horticulturists are experts in
    a. designing outdoor areas for people to use.
    b. how to grow, pick, and ship plants.
    c. artificial insemination.

⑤ Individuals who give advice on the conservation of soil and water are
    a. soil scientists.
    b. agronomists.
    c. Forest Service researchers.

⑥ One of the purposes of processing food is
    a. to decrease the nutritional value.
    b. to study food additives.
    c. to keep it from spoiling.

⑦ A major part of new product development efforts is related to
    a. increased job opportunities.
    b. convenience foods.
    c. the study of food additives.

⑧ Individuals who act as links in the marketing chain between the producers of goods and the retailer are
    a. food technicians.
    b. retail store managers.
    c. wholesalers.

⑨ A landscape architect would be involved in
    a. designing the grounds surrounding an apartment complex.
    b. mowing and fertilizing the grounds of a golf course.
    c. the cultivation of plants.

⑩ Job opportunities in agribusiness have increased because of
    a. the increase in the number of farmers in agriculture.
    b. an increase in the amount of land that is planted.
    c. an increase in the specialization of the processing of food and fiber.

Discussion

① How has the role of education changed for farmers in the last 50 years?

② How does agriculture continue to be an important part of the economy if the number of farmers has declined?

3 List three jobs in the area of plant science.

4 Why do producers today depend on a network of people?

5 Name three topics about which a cooperative extension agent would be able to answer questions.

6 What are some of the tasks of a food scientist?

7 What is agricultural marketing?

8 What challenge does the environment present to producers?

9 Why does it look like careers in the field of agriculture will continue to be important in the future?

10 Why is food processed?

Glossary

A

A horizon   The leached upper portion of a soil profile; the eluvial layer.

Adventitious root   A root growing from a stem or branch and not from root tissue.

Aerator   A mechanical device used to put oxygen into a pond or lake.

Aerial root   A root growing from the stem which is exposed to the air.

Agricultural Education   A high school and/or middle school educational program that consists of three parts—classroom and laboratory activities, supervised agricultural experience (SAE), and FFA.

Agriculture   The industry engaged in the production of plants and animals for food and fiber, the provision of agricultural supplies and services, and the processing, marketing, and distribution of agricultural products.

Alluvial soil   Soil that is transported and deposited by running water.

Amino acids   The building blocks of protein.

Anaerobic bacteria   Bacteria that thrive in the absence of oxygen.

Animal rights   The philosophy that supports the idea that animals have the same rights as humans.

Animal welfare   The philosophy that supports the idea that animals should be well cared for and well treated.

Anther   The part of the stamen on seed-producing plants which develops and contains the pollen.

Antibody   A substance manufactured by an animal's body to combat a disease.

Aquaculture   The growing of aquatic animals for human use.

Aquatic animal   An animal that lives in water.

Aquifer   A geologic formation in rock or sand that contains water.

Artificial insemination (AI)   The placing of sperm in the reproductive tract of the female by means other than the natural breeding process.

Asexual reproduction   The propagating of plants from a single plant.

B

B horizon   The mineral horizon below an A or E horizon.

Bedding plant   Flower and vegetable plants that are planted in beds.

Beef   The meat obtained from cattle. Usually refers to meat from an animal over the age of one year.

Biological control   The control of pests using natural means such as insect predators or weed eating insects.

Biologist   A person who studies living organisms as a career.

Biology   The field of study dealing with living organisms.

Biotechnology   The application of biological and engineering techniques to living organisms.

Blanching   The process of stopping enzyme action in food by dipping the food first in hot water then in cold water.

Border plants   Plants established along the borders of a field to attract insects away from the crop.

Boring Tool   A device used for boring a hole in wood, metal, plastic, or other material.

Bottom Dead Center (BDC)   The position of a piston when it is at the bottom of its stroke.

Bovine somatotropin (BST)   A naturally occurring hormone that aids in stimulating milk production in cows.

Brace   A tool used for turning a bit that bores a hole in wood. It is turned by hand.

Brackish   Water that is neither salt nor fresh but contains a low level of salt.

Breed   Animals having a common origin and distinguishing characteristics.

Broiler   A chicken raised for market that is 2.5 pounds or more. They are usually 6 to 8 weeks old.

Brood pond   A pond used to keep fish that breed and lay eggs.

Brooder   A device used to keep chicks warm.

Bulb   The below-ground bud of a plant.

Bull   A male bovine that has not been castrated.

Butter   A product made from milk fat.

C

C horizon   The mineral horizon or layer, excluding indurated bedrock, that is little affected by soil-forming processes and does not have the properties typical of the A or B horizons.

Caloric content   The amount of energy a food contains.

Cannibalism   The eating or killing of animals by their own kind.

Canning   A food preserving method that uses sealed cans or jars of food that are exposed to high heat to kill microbes.

Canopy   The uppermost growth of crowns of trees.

Carcass   The part of a meat animal remaining after slaughter and the hide, head, feet, and internal organs have been removed.

Carcinogen   Any cancer-causing agent.

Career Development Event   Competitive events conducted as part of FFA. The events help develop career skills.

Cereal   Pertaining to grains or products made from grains.

Cheese   Food made from the consolidated curds that have been separated from milk.

Chisel   A long, flat tool that has a very sharp edge used for cutting.

Cholesterol A fat-soluble substance found in the fat, liver, nervous system, and other areas of an animal's body. It plays an important role in the synthesis of bile, sex hormones, and vitamin D.

Clamp A term used to describe several types of devices used to temporarily hold material together.

Clay A sediment of soft, plastic consistency composed primarily of fine-grained minerals.

Climax vegetation The plants that are the last to appear as a result of natural succession.

Closed-loop system A system of tanks where fish are raised. Oxygen is provided and the water is filtered.

Cocoon The covering that surrounds an insect pupa.

Cold frame An enclosed unheated frame used for growing plants.

Combine A self-propelled or tractor-drawn machine that cuts, threshes, and cleans the standing crop while moving across the field.

Combustion The chemical process of oxidation that produces heat.

Commercial fertilizer Plant nutrients derived from processing involving chemical mixtures and sold to producers.

Commodity A transportable resource product with commercial value; all resource products that are articles of commerce.

Composting The process of piling organic matter together so it will undergo chemical changes that make the matter useable for fertilizer.

Compression ratio A comparison of the volume of an engine cylinder at top dead center (TDC) to the volume of the cylinder at bottom dead center (BDC).

Concentrate A feed high in carbohydrates and low in fiber.

Confinement operation Operating systems where animals are confined in buildings or cages.

Conifer An evergreen tree that produces cones.

Consumer A person who buys goods or services from other people.

Controlled environment An environment for animals that is kept at the correct temperature and other conditions to maximize animal comfort.

Cotton gin A machine used to separate the cotton seed from the lint.

Cover crop Crops established for the purpose of adding organic matter to the soil. They are usually planted in the winter and plowed under in the spring.

Cow A female bovine that has had a calf.

Crop residue Stalks and other debris left in the field from last year's crop.

Crop rotation Planting a different crop than was planted the last year.

Crossbreed An animal with parents of different breeds.

Cross-pollination Pollination that is the result of gametes from different plants.

Crustacean Aquatic animals with a rigid outer covering, jointed appendages, and gills.

Cultivated plant Any plant that has been planted, tended, harvested, and improved.

Curd A milk product consisting of casein and fat; the coagulated part of milk.

Customary Measuring System A measuring system used in everyday life in the United States. Inches, feet, and yards are common measurements in the customary system.

Cut flower A flower that has been harvested by cutting it off with the stem attached.

Cutting A part of a plant that has been cut off for the purpose of propagating a new plant.

Cycle The complete sequence of events that occurs within a combustion chamber of an internal combustion engine.

D

Dairy Pertaining to products that are derived from milk. Also, a farm where cows are kept for milk production.

Dam breed A breed of animal that makes good mothers that are used in crossbreeding programs.

Dam The mother of an animal.

Debeaking The removal of the beak from a baby chicken.

Deciduous Refers to a plant that loses its leaves every year.

Diet The amount of food taken in during a 24-hour period. The term can be used in referring to either human food or animal feed.

Disease Any departure from the normal health of a plant or animal.

Disinfectant A substance that kills pathogens (germs).

Dissolved oxygen Oxygen that is dissolved in water and is usable by animals that breath through gills.

Division A means of plant propagation that involves cutting crowns, rhizomes, tubers, or roots into sections.

Docile Refers to an animal that is gentle in nature.

Docking The removal of an animal's tail. This is done to help keep the animal clean or to prevent tail biting.

Dressed Refers to an animal that has been slaughtered and the carcass cleaned.

Drug residue Any amount of a drug that is left in an animal's body.

Drying A process of food preservation where moisture is removed from the food.

E

E horizon A mineral horizon, mainly a residual concentration of sand and silt high in content of resistant minerals as a result of the loss of silicate clay, iron, aluminum, or a combination of these.

E. coli bacteria   A bacteria that lives in the colon of mammals. It can cause a variety of diseases.

Ecology   The totality or pattern of the interrelationship of organisms and their environment, and the science that is concerned with that interrelationship. The subject is often categorized into plant, animal, marine, terrestrial, freshwater, desert, tundra, and tropical ecology.

Ecosystem   The entire system of life and its environmental and geographical factors that influence all life, including the plants, animals, and the environmental factors.

Ectothermic   Refers to an animal that gets its body temperature from its environment.

Electronic Measurers   Devices that measure distances electronically.

Endothermic   Refers to an animal that generates its own body temperature internally.

Engine   A device or machine that can produce power on its own independent from an external source of energy.

Engine timing   The correct sequence of ignition, piston movement, and valve opening and closing that occurs in a properly running engine.

Environment   The sum total of all the external conditions that may act on an organism.

Environmental Protection Agency (EPA)   An agency of the federal government that oversees environmental concerns.

Enzyme   A substance that speeds up or stimulates a chemical reaction.

Erosion   The wearing away of the soil by water or wind action.

Exhaust stroke   The movement of the piston within a cylinder that expels exhaust materials.

Exotic plant   A plant that is out of the ordinary.

Experiment station   Any of the USDA research facilities associated with state agricultural universities, where new ways of farming, ranching, or rural living are officially tested.

Export   Shipment of commodities to foreign countries.

F

Farrow   The process of giving birth to pigs.

Farrowing crate   A cage or crate in which a sow is placed for farrowing. It keeps her from crushing the pigs.

Feedlot   A lot where cattle are kept for fattening for market.

Fermentation   The processing of food by the use of yeasts, molds, or bacteria.

Fertilizer   Any organic or inorganic material added to soil or water to provide plant nutrients and to increase the growth, yield, quantity, or nutritive value of the plants grown therein.

Fibrous root   A root system which is composed of many fine, branched roots.

Fingerling   A fish from 3 to 6 inches long that is used for stocking ponds or lakes.

Fish slurry A material made from the entrails, heads, bones, etc. of fish. It is a rich source of plant nutrients.

Flock A group of poultry or sheep.

Flood plain The area along a river or body of water that is covered in water by periodic flooding.

Floriculture The cultivation of plants for their flowers.

Florist A person who deals in arranging and selling flowers.

Flour The fine-ground product obtained in the commercial milling of wheat and other grains which consists essentially of the starch and gluten of the endo-sperm. It is used for baking bread, pastry, cakes, etc.

Foliage plant A plant that is grown for its attractive foliage.

Food and Drug Administration (FDA) An agency of the federal government that regulates food and drugs that are sold to consumers.

Food chain The transfer of food energy from the initial source in plants through a series of organisms by repeated eating and being eaten.

Forestry The science of growing trees.

Four-stroke cycle engine An internal combustion engine that requires four complete movements of the piston to complete the cycle.

Free range Allowing animals to live in an area unrestricted by a fence.

Freeze-drying A food preserving method that uses a vacuum to remove moisture from frozen foods.

Fresh water Water that contains no or very little salt content.

Fruit The matured ovary of a flower and its contents including any external part which is an integral portion of it.

Fry Small, newly hatched fish.

Fumigate The destruction of pathogens by the use of liquids or solids that release a vapor.

Fungi Plantlike organisms that have no chlorophyll; they get their nourishment from living or decaying organic matter. Plural of fungus.

G

Gaited The definite rhythmic movement of a horse such as trot, canter, pace, etc.

Game bird A bird that is hunted for food.

Gene The simplest unit of inheritance. It is composed of DNA.

Genetic engineering The altering of the genetic components of an organism by human intervention.

Geology The study of the Earth, earthy materials, their history, processes, and products. This includes water, air, minerals, and rocks.

Germination The emerging of a plant from a seed.

Glacier A slow-moving mass of ice.

Greenhouse Any of several different types of heated, glass- or plastic-covered structures used for the growing of plants.

Groundwater Water within the ground that supplies wells and springs.

Growing season   The number of days from the average date of the last freeze in the spring to the average date of the first freeze in the fall; the period in which most plants grow.

H

Hammer   A tool used for driving nails.

Hand Tool   Tools that are used and powered by humans.

Hardwood   Wood from a deciduous tree.

Heavy soil   Generally, a clayey soil in contrast with a sandy one. The reference is to resistance and need for greater power in pulling a plow rather than to actual weight or specific gravity which is less for dry clay than for dry sand.

Heifer   A female bovine that has not had a calf.

Herbicide   A substance that is used to kill plants.

Heterosis   The degree of superiority of a crossbred animal over what would be expected of its purebred parents. Also known as hybrid vigor.

Hive   A box used to hive a swarm of bees.

Homogenized milk   Milk that has been processed to blend the butter fat.

Hormone   A chemical substance produced by an organism that causes a specific effect on the animal or a portion of its systems.

Horticulture   The branch of agriculture that deals with the growing of fruits, flowers, vegetables, or ornamental plants.

House plant   An ornamental plant that is grown for its beauty and kept in the house.

Humus   Organic matter in the soil.

Hunters and gatherers   People who obtain their food by gathering wild plants and animals.

Hutch   A cage and shelter to house rabbits.

Hybrid vigor   *See* heterosis.

Hydrological cycle   The cycle of water evaporating and condensing and beginning the cycle over again.

I

Import   Goods that are brought into this country from another country.

Incubator   A device that controls the temperature and relative humidity for the hatching of eggs.

Infection   The invasion of an organism's tissues by a disease-causing agent.

Inorganic soil   A mineral soil; a soil in which the solid matter is dominantly rock minerals in contrast to organic soils, such as peats and mucks.

Insecticide   A substance used to kill insects.

Intake stroke   The movement of the piston within a cylinder that draws fuel into the cylinder.

Integrated pest management   A method of controlling pests that uses a variety of means.

Interiorscaping The use of plants inside a building to produce a pleasing effect.

Internal Combustion Engine A machine or device that is capable of converting heat energy into mechanical power. The heat is produced inside of the machine.

Irrigation The artificial application of water to soil for the purpose of increasing plant production.

J

Jerky A meat product made by drying.

K

Kerf The gap cut into a board or other material by a saw.

Koi Fish of the carp family that are raised because of their colorful appearance.

L

Lagoon A pond used for disposing of animal manure.

Land-grant university Any of the state colleges and universities started from federal government grants of land to each state to encourage further practical education in agriculture, homemaking, and the mechanical arts. The mission of these universities is to conduct programs in teaching, service, and research.

Landscape architect A person trained in the art and science of arranging land and objects on it for human enjoyment and use.

Landscaping To beautify the terrain with the planting of trees, shrubs, and plants.

Landscaping cloth A cloth used to prevent weeds from growing among landscape plants.

Larvae The immature insect as it hatches from the egg.

Layer A hen that is raised for producing eggs.

Legume A family of plants that with the aid of symbiotic bacteria converts nitrogen from the air to build up nitrogen in the soil.

Level A tool that uses a bubble inside an enclosed glass capsule of mineral spirits to indicate levelness.

Life cycle The changes in life form that an organism undergoes.

Litter A group of young pigs that were born from the same mother. Also the absorbent material spread on the floor of an animal's pen or house.

M

Mad Cow Disease A disease in cattle that is caused by a type of protein that attacks the nervous system. Mad cow Disease can be transmitted to humans if they eat poorly cooked beef from an animal with the disease.

Magneto ignition system An engine electrical system in which a spark is generated from a magnetic field created by magnets.

Marking Tool A tool to mark a line on a board or other material.

Mastitis A disease of the mammary glands that is usually caused by an injury.

Measuring Device Any tool used to measure.

Measuring Tape A device that measures linear distances. It is flexible and rolls up into a case.

Media Soil or soil-like material in which plants grow.

Meteorology The study of the physical processes which occur in the atmosphere and of the related processes of the lithosphere and hydrosphere.

Metric Measuring System A measuring system based on a decimal system, with the meter being the base for measurement. Measuring devices in the metric system are marked as meters, decimeters, centimeters and millimeters.

Microorganism An organism so small that it cannot be seen clearly without the use of a microscope; a microscopic or submicroscopic organism.

Mineral A chemical compound or element of inorganic origin.

Motor A device or machine that produces power from the energy supplied by an external source, such as electricity.

Mule An animal produced by the mating of a jackass and a mare.

Mutton The meat of a mature sheep.

N

National FFA Organization A program conducted as a part of Agricultural Education that centers on leadership development and personal growth.

Natural resource Any resource that grows or occurs naturally.

Natural toxins Poisons released by plants as a natural defense against insects.

Navigable waterway A stream or body of water that is large enough to allow the passage of boats or ships.

Nectar The sweet substance from flowers. Bees make honey from this material.

Nematode Microscopic, wormlike, transparent organism that can attack plant roots or stems to cause stunted or unhealthy growth.

Node The place on a stem which normally bears a leaf or whorl of leaves.

Nodule A root formation on certain leguminous plants produced by invasion of symbiotic, nitrogen-fixing bacteria. The bacteria furnish the plant with fixed nitrogen compounds and receive nutrient plant juices like carbohydrates.

Nonpoint source Pollution that occurs from several sources rather than a single source.

Nonselective herbicide A substance that kills all plants it contacts.

Nutrient An element or compound in a soil which is essential for the growth of a plant.

O

O horizon Organic horizons above mineral soil.

Orchard A field used for growing fruit trees.

Organic fertilizer Plant nutrients derived from organic sources such as compost, manure, etc.

Organic matter Matter found in or produced by living animals and plants which contains carbon, hydrogen, oxygen, and often nitrogen and sulfur.

Organic soil A soil having a high content of organic soil materials.

Ornamental plant Any plant produced for its beauty.

Ovary The female organ that produces eggs and female hormones.

P

Parasite An organism that lives off of or at the expense of another organism.

Pasteurization The process of killing microbes in milk by the use of heat.

Pasture An area of grass maintained for animal grazing.

Pesticide Any substance used to kill pests.

Pesticide residues Tiny amounts of pesticides that are left on produce after harvesting.

pH A numerical measure of acidity or hydrogen ion activity of a substance such as food or soil. The neutral point is pH 7.0. All pH values below 7.0 are acid and all above 7.0 are alkaline.

Pharmaceutical Any substance used to enhance the health of humans or animals.

Phloem The tissue of a plant that transports food.

Photoperiodic Plants that react to specific periods of light.

Photosynthesis The process in plants that converts light energy to chemical energy.

Pickling A method of preserving food that uses an acid such as vinegar as a preservant.

Pistil The female element of a flower.

Plane A hand tool used to smooth boards.

Plumb Bob A tool consisting of a weight on the end of a string that indicates whether a material is plum (vertically level).

Point source Pollution that occurs from a single source.

Pollination The transfer of pollen from the anther to the stigma of a flower.

Pollution Substances in a body of water, air, soil, etc., that impair the usefulness of it.

Pork The meat from pigs.

Potted plant A plant that lives its entire life in a pot.

Poultry   Domestic fowls that are raised for meat, eggs, or feathers.

Power   A source of energy or a means of completing work.

Power Stroke   The movement of the piston within a cylinder that creats power.

Power Tool   A tool that has an outside power source such as electricity.

Predator   An animal that eats other animals.

Predator insects   Insects that eat other insects.

Proficiency Award   An award given for excellence in Supervised Agricultural Experience Programs (SAEP).

Propagation   Increasing the number of plants by causing them to reproduce.

Protozoa   A group of one-celled organisms that generally do not contain chlorophyll; sometimes classified as one-celled animals.

Pruning   The removing of limbs or other plant parts to shape the plant.

Pupa   The stage of an insect's life between the larva and the adult stage.

Purebred   An animal belonging to one of the recognized breeds. The animal's ancestry are all of that breed.

R

Ration   All the feed an animal receives during a 24-hour period.

Refrigeration   Artificial cooling, either by the application of ice or by utilizing the principle of the latent heat of evaporation.

Registration   The process of certifying that an animal is a purebred of a particular breed.

Renewable resource   A resource that can be replaced such as trees that are grown as a crop.

Rennet   A substance derived from the stomachs of calves and used in the cheese process.

Resistant   Designating a plant or animal that can resist disease.

Rhizobia   Bacteria that live symbiotically in roots of legumes and fix nitrogen from the air.

Rhizosphere   The root zone, the area where microorganisms are most active in increasing the availability of nutrients for plants.

Ripping   The act of cutting a board lengthwise (with the grain).

Roe   A mass of fish eggs.

Root cutting   A cutting taken from a plant root for the purpose of growing a new plant.

Rooting hormone   A substance applied to a plant or root to stimulate growth.

Roughage   A feed that is high in fiber content and low in carbohydrate content.

Ruler   A short tool used to measure lines.

S

Salting   A food preserving method that uses salt as a preservant.

Sand   Any of the small, loose, granular fragments which are the remains of disintegrated rocks. It may contain a great variety of minerals and rocks, but the most common mineral is quartz.

Scavenging   The process of moving the exhaust gases from the cylinder of an engine.

Screwdriver   A tool used for turning a screw in wood, metal, plastic or other material.

Sea island cotton   A woody plant grown as an annual for its long staple lint cotton used in fine cotton fabrics. Native to tropical America.

Seed   The embryo of a plant.

Seedling   A plant that has newly emerged from the seed.

Selective breeding   Choosing the most desirable plants or animals for breeding in order to develop superior plants or animals.

Selective herbicide   A substance that will kill some plants and have little or no effect on other plants.

Self-pollinating   A plant that produces and uses its own pollen for reproduction.

Semen   The fluid from a male animal that contains the sperm.

Sexual reproduction   Reproduction through the means of the uniting of gametes.

Shade cloth   A cloth used to cover plants to reduce the amount of sunlight.

Shrub   A plant that has several woody stems and is low growing.

Silage   A feed made from chopped and fermented corn.

Silt   Small, mineral soil particles ranging in diameter from 0.05 to 0.002 mm in the USDA system or 0.02 to 0.002 mm in the International system.

Sire   The father of an animal.

Sire breed   Breeds of animals that are used primarily as sires in a crossbreeding system.

Skim milk   Milk that has had almost all of the milkfat removed.

Smith-Hughes Act   A United States federal law passed by Congress in 1917 which gives federal aid for vocational education in the secondary schools.

Softwood   Wood from a conifer tree.

Soil   The mineral and organic surface of the earth capable of supporting upland plants.

Soil amendment   Any material added to the soil that improves it.

Soil texture   The coarseness or fineness of soil.

Spawn   The process of a fish laying eggs.

Squab   The meat from a pigeon.

Square   A tool used for laying out lines and angles.

Stamen   The male part of a flower that bears the pollen.

Stigma   The female part of the flower that receives the pollen.

Stocker   A beef animal that weighs around 700 to 800 pounds that needs additional growth before entering the feedlot.

Strip cup   A cup used to catch the first milk milked from a cow.

Stroke   The up or down distance traveled by a piston in an internal combustion engine.

Succession   The replacement of one type of plants for another in the natural progression of a forested area.

Supervised Agricultural Experience   An after school educational program connected with a high school Agricultural Education Program that is aimed at gaining practical experience in some area of agriculture.

Surface water   Water that lies on the surface of the ground.

Symbiotic relationship   A relationship between two different types of organisms that is beneficial to both of them.

T

Taproot   The main descending root of a plant from which branch roots grow.

Teat   The part of the female mammary system from which the young receives milk.

Terminal tip   The end of a branch, stem, or root where growth takes place.

Tissue culture   The process of growing a new plant from plant cells.

Tolerant   The ability of a plant or animal to resist a pest or disease.

Top Dead Center (TDC)   The position of a piston when it is at the top of its stroke.

Topiary   The shaping of plants into shapes that are not natural such as into the shape of animals.

Trade balance   The balance of imports and exports of a country.

Tree farm   A farm that grows trees for commercial use such as pulp, lumber, Christmas trees, etc.

Two-stroke cycle engine   An internal combustion engine that requires two complete movements of the piston to complete the cycle.

U

Udder   The exterior portion of a females mammal's mammary system.

Upland cotton   A tropical, woody herb grown as an annual for its fiber, the source of much of the commercial cotton in the United States.

USDA   The United States Department of Agriculture, a department in the Federal Government that oversees matters concerning agriculture.

V

Veal   Meat from a calf that is less than 3 months old.

Vegetable   The edible part of an herbaceous plant.

Vocational Agriculture   A program designed for rural youths in which they studied agriculture. The program has been revised and is now called Agricultural Education.

Volcanic ash   The fine particles from volcanic eruptions which, in thick deposits, are the parent material of soil.

W

Wean   To take a young animal away from its mother when it is mature enough to eat on its own.

Weed   Any plant that is growing where it is not wanted.

Wetlands   An area that stays wet for most of the year.

Whey   The watery portion of milk that remains after the curd and cream have been removed.

Whorled   Three or more leaves, twigs, etc., that are arranged in a circle at one point on a plant.

Wilderness area   An area set aside by the government to preserve the wild state of the area. Access is limited.

Wildlife habitat   The environment in which wildlife lives.

Windbreaks   Trees that are planted to block the wind from homes, roads, crops, etc.

X

Xylem   Plant tissue that transports water.

Y

Yogurt   A semisolid fermented milk food product.

Index

O